Boston Studies in the Philosophy and History of Science

Volume 314

Series editors

Alisa Bokulich, Department of Philosophy, Boston University, Boston, MA, USA
Robert S. Cohen, Boston University, Watertown, MA, USA
Jürgen Renn, Max Planck Institute for the History of Science, Berlin, Germany
Kostas Gavroglu, University of Athens, Athens, Greece

The series *Boston Studies in the Philosophy and History of Science* was conceived in the broadest framework of interdisciplinary and international concerns. Natural scientists, mathematicians, social scientists and philosophers have contributed to the series, as have historians and sociologists of science, linguists, psychologists, physicians, and literary critics.

The series has been able to include works by authors from many other countries around the world.

The editors believe that the history and philosophy of science should itself be scientific, self-consciously critical, humane as well as rational, sceptical and undogmatic while also receptive to discussion of first principles. One of the aims of Boston Studies, therefore, is to develop collaboration among scientists, historians and philosophers.

Boston Studies in the Philosophy and History of Science looks into and reflects on interactions between epistemological and historical dimensions in an effort to understand the scientific enterprise from every viewpoint.

More information about this series at http://www.springer.com/series/5710

Pierre Maurice Marie Duhem

The Electric Theories
of J. Clerk Maxwell

A Historical and Critical Study

Translated by Alan Aversa

 Springer

Pierre Maurice Marie Duhem
Cabrespine
France

Translated by Alan Aversa

Pierre Maurice Marie Duhem is deceased

ISSN 0068-0346 ISSN 2214-7942 (electronic)
Boston Studies in the Philosophy and History of Science
ISBN 978-3-319-36785-9 ISBN 978-3-319-18515-6 (eBook)
DOI 10.1007/978-3-319-18515-6

Printed on acid-free paper

Springer International Publishing AG Switzerland is part of Springer Science+Business Media
(www.springer.com)

To my wife, who made this translation possible, and to the Holy Trinity, Who makes all things possible

Foreword

Pierre Maurice Marie Duhem[1] (1861–1916)—an accomplished physicist,[2] philosopher of physics,[3] and historian of physics[4]—ranked first in his class at the *École Normale Supérieure* (Jaki 1984, 36) France's most prestigious university, and first on the national physics *concours* exam for *agrégation* in 1885 (Chervel 2011). In his third year at the *École Normale,* he was the first student ever in France's *grandes ecoles* to present himself for the doctor's degree (Jaki 1984, 47). The thesis, later reprinted as *The Thermodynamic Potential and its Applications to Chemical Mechanics and to the Study of Electrical Phenomena,*[5] was rejected because it disproved the "principle of maximum work" of Berthelot, who had great influence over French academic politics. Undiscouraged, Duhem presented a second thesis, *On Magnetization by Induction* (Duhem 1988), this time in mathematics. It was accepted by a committee including Poincaré and Tannery. Tannery cataloged the thesis under the title: *A Novel Theory of Magnetization by Induction*

---

[1]Jaki (1984); Maiocchi (1985); Miller (1970); Ariew (2011); Duhem (1936)

[2]Jaki (1984, 259–317). For a physicist's perspective on Duhem (1902), see de Moura and Sarmento (2013).

[3]Jaki (1984, 319–373). Duhem's greatest, most well-known work in the philosophy of physics is Duhem (1906), translated as Duhem (1991); it even influenced Einstein (Howard 1990). For a philosophical perspective on Duhem (1902), see Ariew and Barker (1986). For how Duhem's philosophy of physics relates to his cosmological, thermodynamical, and even religious views, see Kragh (2008).

[4]Jaki (1984, 375–436). Duhem (1990b) and Duhem (1990a) are his own summaries of his philosophy and history of physics for his candidacy in the Académie des Sciences, originally published as Duhem (1913, 151–157) and Duhem (1913, 158–169), respectively. Duhem's ten volume *Système du monde* (Duhem 1913–1959), partially translated as Duhem (1985), initiated the field of the history of medieval physics. It also demonstrates Duhem's "continuity thesis" of scientific development (Hannam 2009). For how Duhem's historiography influenced his epistemology, see Bordoni (2013b).

[5]*Le potentiel thermodynamique et ses applications à la mécanique chimique et à l'étude des phénomènes électriques* (Duhem 1886)

Founded on Thermodynamics[6]; by including the term "thermodynamics," he emphasized that this thesis is very similar in content to Duhem's rejected thesis.

The subject of Duhem's theses reflects his grand vision for physics: to subject all branches of physics—mechanics, chemistry, electromagnetism, etc.—to thermodynamic first principles.[7] Drawing inspiration from the "energetics program" (Rankine 1855) of Rankine (Parkinson 2008), Duhem subjected mechanics and chemistry to thermodynamic first principles in works such as his *Commentary on the Principles of Thermodynamics* (Duhem 2011), one of Duhem's few scientific works translated into English, and the *Treatise on Energetics or General Thermodynamics* (Duhem 1911), which Duhem considered his greatest scientific achievement.[8]

Duhem's Philosophy of Physics

Duhem was a moderate realist (Brenner et al. 2011, 7–12) who argued that physical theories are classifications of experimental laws. This is a key aspect of Duhem's philosophy of physics:

> A physical theory…is an abstract system whose aim is to *summarize* and *classify logically* a group of experimental laws without claiming to explain these laws.[9]

Duhem (1991 19) gives a more specific definition of physical theory in terms of the "abstract system" of mathematics:

> A physical theory is not an explanation. It is a system of mathematical propositions, deduced from a small number of principles, which aim to represent as simply, as completely, and as exactly as possible a set of experimental laws.[10]

Just as there are many ways to classify seashells or bodily organs, so there are also many ways to classify classification physical laws; and just as classifications *per se* do not explain what they classify, so also physical theories do not explain physical laws. Physical theories are not, as Newton thought about his theory of

---

[6]*Théorie nouvelle de l'aimantation par influence fondée sur la thermodynamique* (cf. Jaki 1984, 79).

[7]For how Duhem partially accomplished this task, see Bordoni (2012a,b,c, 2013a). Needham (2013) is a review of Bordoni (2012a).

[8]For a translation of the introduction of Duhem's *Treatise*, see Maugin (2014, 172–175).

[9]Duhem (1991, 7), a translation of Duhem (1906, 3):
> Une théorie physique. est un systéme abstrait qui a pour but de *résumer* et de *classer logiquement* un ensemble de lois expérimentales, sans prétendre expliquer ceslois.

[10]Original from Duhem (1906, 24):
> Une théorie physique n'est pas une explication. C'est un systeme de propositions mathématiques, déduites d'un petit nombre de principes, qui ont pour but de représenter aussi simplement, aussi completement et aussi exactement que possible, un ensemble de lois expérimentales.

gravitation or Ampère about his force law, "uniquely deduced from experience;" other theories (e.g., Einstein's theory of gravitation) can equally, if not better, "save the phenomena save the phenomena"[11] of experience. In the history of physics, Duhem sees—from as far back as Aristotle to the present day—a long, steady, and continuous process asymptotically approaching the best, "natural classification."

All of Duhem's philosophy of physics—the under-determination of theory by fact, confirmation holism,[12] the strict separation between physics and metaphysics,[13] and the continuity of scientific development—is rooted in his understanding of physical theory as a classification.

One reason Duhem preferred Helmholtz's electromagnetic theory over that of others,[14] in addition to its being in "continuity with tradition," is because of what Buchwald calls Helmholtz's "taxonomy of interactions" between "laboratory objects."[15] Helmholtz's approach to electromagnetism was to classify the unique interaction energies between the various combinations of charged and current-carrying "laboratory objects." Thus, Helmholtz explicitly classified experimental laws, forming a true theory in the Duhemian sense.

Reception of Duhem's Physics

Lorentz (1926, 65), an article on Maxwell's electromagnetic theory, cites Duhem (1902), of which the present work is the translation, and classifies Duhem's treatment of electrodynamics under the heading "43. *Thermodynamische Behandlung*" ("Thermodynamic Treatment"), saying: "*P. Duhem* represents, in particular, the thermodynamic viewpoint."[16] He cites Duhem's *Lessons on Electricity and Magnetism* (Duhem 1891–1892), his accepted thesis (Duhem 1888), and his article in the American Journal of Mathematics (Duhem 1895b); however, Lorentz thought it would take him too far afield to discuss Duhem's treatment in detail.

Louis Roy, a student of Boussinesq and Professor of physics at the University of Toulouse (Jaki 1984, 298), promoted Helmholtz-Duhem electrodynamics in great detail in a book (Roy 1923a) and several articles (Roy 1915, 1918, 1923b). He writes[17]:

---

[11]Duhem (1908), translated as Duhem (1969).

[12]i.e., that there are no "crucial experiments crucial experiment;" cf. the related Duhem-Quine thesis: Ariew (1984).

[13]Duhem (1893), translated as Duhem (1996, 29–49).

[14]See this volume p. xx.

[15]Buchwald (1994, 11–12); cf. Buchwald's article "Electrodynamics in Context: Object States, Laboratory Object States, and Anti-Romanticism" (Cahan 1993, 334–373).

[16]"Den thermodynamischen Standpunkt hat insbesondere *P. Duhem* vertreten." (Lorentz 1926, 140-1).

[17]Roy (1923a, 7; 87), translated in O'Rahilly (1938, 178).

Maxwell kept his eyes fixed on his object, which was to establish a theory inclusive of electrical and optical phenomena; unfortunately none of the paths he successively followed could lead him thereto. Then, when logic barred the way, he evaded the inconvenient obstacle by a flagrant fault of reasoning or calculation, certain that his objective was true. ... The best way of recording our admiration for such a genius, is to reformulate his work with the help of the ordinary laws of logic. ... An excessive admiration for Maxwell's work has led many physicists to the view that it does not matter whether a theory is logical or absurd, all it is required to do is to suggest experiments. ... A day will come, I am certain, when it will be recognised ... that above all the object of a theory is to bring classification and order into the chaos of facts shown by experience. Then it will be acknowledged that Helmholtz's electrodynamics is a fine work and that I did well to adhere to it. Logic can be patient, for it is eternal.

[The Helmholtz-Duhem exposition is] the only real demonstration of Maxwell's equations which has hitherto been given.

In 1938, Alfred O'Rahilly devoted a whole chapter of his two-volume *Electromagnetic Theory: A Critical Examination of Fundamentals* to Helmholtz-Duhem theory (O'Rahilly 1938, 161–180), citing Duhem (1902) copiously.[18] He concludes (O'Rahilly 1938, 177):

We have just shown that it is impossible to admit that Helmholtz's theory, as just expounded, really re-establishes the tradition of writers like Weber and C. Neumann, not to speak of the contemporary electron theory. Nevertheless Duhem's work is of permanent value, and his protest against the complaisant acceptance of contradictory standpoints is still apposite.

That it "is still apposite" is evidenced by the fact that, very recently, Maugin (2014, 104–107) discusses Helmholtz-Duhem theory in the context of "incorporating electricity and magnetism, including nonlinear dissipative effects such as hysteresis, in his broad energetic view." (Maugin 2014, 104). Duhem never accomplished this in his great *Treatise*; thus, it remains an open problem for young physicists to tackle (cf. Wipf 2011).

Note on the Translation

Page numbers in [•] brackets, refer to the pages of the original (Duhem 1902). Page numbers in [•] brackets in the citations in the footnotes refer to the page numbers of the English version of the citation.

Sierra Vista, Arizona   Alan Aversa
March 2015

---

[18]O'Rahilly (1938, 36; 79–80; 83; 90; 95–96; 177; 182; 210).

Acknowledgments

In September 2013, I asked Stefano Bordoni what would be a good journal to publish Duhem (1895a), but he suggested that it would be better to translate all of Duhem (1902). He was certainly right. I would also like to thank Stanley L. Jaki[19] for his tireless work in promoting Duhem's physics and Ryan Vilbig for introducing me to Alfred O'Rahilly,[20] an advocate of alternative theories of electrodynamics, such as Weber's theory, Ritz's ballistic theory of light, and the Helmholtz-Duhem theory.

---

[19]†2009, *requiescat in pace.* http://www.sljaki.com/.
[20]http://humphrysfamilytree.com/ORahilly/alfred.html.

Contents

Chapter 1
Introduction

In the middle of this century,[1] electrodynamics seemed established in all its essential parts. Awaken by the experience of Oersted, the genius of Ampère had created and brought to a high degree of perfection the study of forces acting between two currents or between a current and a magnet; Arago had discovered magnetization by currents; Faraday had highlighted the phenomena of electrodynamic induction and electromagnetic induction; Lenz had compared the sense of electromotive actions [2] of currents to the sense of their ponderomotive actions. This comparison provided to F.E. Neumann the starting point of a theory of induction. W. Weber proposed this theory, in relying on hypotheses of the general laws of electric forces. Finally, H. Helmholtz, then W. Thomson, attempted to pass from the laws of Ampère to the laws discovered by F.E. Neumann and W. Weber, taking the newly asserted law of the conservation of energy as an intermediary principle.

Only two objects seemed to offer themselves to the study of the physicist eager to work in the progress of electrodynamics and electromagnetism.

The first of these objects was the development of the consequences implicitly contained in the principles that had been posed. To pursue this object, the geometers employed the resources of their analysis; experimenters began implementing their most accurate measurement methods; industrials lavished their inventive ingenuity; and, soon, the study of electricity became the richest and largest chapter of all of physics.

The second of these objects, of a more speculative and more philosophical nature, was the reduction to a common law of the principles of electrostatics and electrodynamics. Ampère himself had proposed it to the efforts of physicists. He said[2]:

> It is therefore completely demonstrated that one cannot make sense of the phenomena produced by the action of two voltaic conductors, assuming that electric molecules acting

---

[1][The 19th century].

[2]Ampère. *Théorie mathématique des phénomènes électrodynamiques uniquement déduite de l'expérience*, Paris, 1826. Deuxième édition (Paris, 1883), pp. 96 et sqq. [English translation: Ampère (2015)].

© Springer International Publishing Switzerland 2015
P.M.M. Duhem, *The Electric Theories of J. Clerk Maxwell*,
Boston Studies in the Philosophy and History of Science 314,
DOI 10.1007/978-3-319-18515-6_1

inversely to the square of the distance were distributed on the conductive wires in such a way so as to remain fluxed and can, therefore, be viewed as invariably linked among themselves. It must be concluded that these phenomena are due to that the two electrical fluids roam continually the conductive wires with an extremely fast movement, [3] meeting and parting alternately in the gaps of the particles of these wires...

It is only in the case where one assumes the electric molecules at rest in the body, where they manifest their presence through the attractions or repulsions produced by them between these bodies, that it is shown that a uniformly accelerated movement neither can result from the forces exerted by the electric molecules in this state of rest nor depend only on their mutual distances. When it is assumed, instead, that, put in motion in the wires by the action of the battery, they are continually changing place, gather at every moment in neutral fluid, separate again, and will meet in other fluid molecules of the opposite nature, it is not more contradictory to admit that from the actions in inverse ratio of the squares of the distances which exert on each molecule, a force can arise between two elements of conductive wires which depends not only on their distance, but also on the directions of the two elements whereby electric molecules move, gather in the molecules of the opposite species, and separate the next moment to unite with others...

If it were possible, starting from this consideration, to prove that the mutual action of the two elements is, indeed, proportional to the formula by which I represented it, this explanation of the fundamental fact of the theory of electrodynamic phenomena should obviously be preferred to any other...

To the question that Ampère only posed, Gauss[3] formulates a response that he did not publish: the mutual repulsive action of two electrical charges does not only depend on their distance, but also on the speed of relative motion of the one with respect to the other; when two charges are at relative rest, this action reduces to the force inversely proportional to the square of the distance, known since Coulomb; when, on the contrary, two conductive wires are, the one and the other, the seats of two electrical currents leading in opposite directions, with equal [4] speed, one the positive electricity and the other the negative electricity, these two wires attract one another according to Ampère's law.

Gauss merely put on paper a formula that answered the question of Ampère; his illustrious pupil, W. Weber,[4] imagined a similar formula and deduced all the consequences. According to Weber, the mutual action of two electrical charges depends not only on their distance, but also on the first two derivatives of this distance with respect to time. Reproducing Coulomb's law when applied to electrostatic phenomena, the formula of Weber indicates that both current elements attract according to the formula of Ampère. In addition, it provides a complete mathematical theory of electrodynamic induction, a theory consistent at all points to that which F.E. Neumann discovered at the same time, inspired by the methods of Ampère.

The doctrine of Weber was, first of all, great; most physicists considered, according to the words of Ampère, that "this explanation of the fundamental fact of the theory of electrodynamic phenomena should be preferred to any other."

---

[3]C.F. Gauss, *Werke*, Bd. V, p. 616.

[4]Weber, *Elektrodynamische Maassbestimmungen*, I, Leipzig, 1846.

However, this doctrine did not justify the hopes it first raised, although G. Kirchhoff[5] deduced, for induction within conductors of finite extent in all dimensions, a theory that served as a precursor to the research of Helmholtz, it did not lead to the discovery of any new fact; and, little by little, desperate by the sterility of the speculations regarding the actions that carry electric charges in motion, physicists diverted their attention, which could not be brought back by the hypotheses of B. Riemann, nor by the researches of R. Clausius. [5]

So, electrodynamics appeared in 1860 as a vast country whose daring explorers recognized all the frontiers; the exact scope of the region seemed known. It only remained to study carefully each of its provinces and exploit the riches it promised for industry.

However, in 1861, to this science that seemed so completely master of its domain, a new and vast area was opened; and so one could believe, many think today, that this sudden extension should not only increase electrodynamics, but also upset parts of this doctrine that are regarded as established in an almost final manner.

This revolution was the work of a Scottish physicist, James Clerk Maxwell.

Taking up and developing the old ideas of Aepinus and Cavendish, Faraday created, besides the electrostatics of conductive bodies, the electrostatics of the insulating body or, in the words he introduced in physics, *dielectric* bodies; but no one had taken these bodies into in account in the speculations of electrodynamics. Maxwell created the electrodynamics of the dielectric body. He imagined that the properties of dielectrics, at any given time, depended not only on the polarization of this body at this moment, but also on the speed with which the polarization varies from one moment to the next; he assumed that this speed causes ponderomotive and electromotive forces similar to those that cause the flow of electricity. To the *conduction current* he compared the *polarization current* or, in his words, the *displacement current*.

Not only do displacement currents exert, in conductive bodies, inducing actions similar to those of the conduction current, but also the electromotive forces of these two kinds of current, giving rise to a current in a conductive body, polarize the dielectrics in which they act. The equations, which derived from these hypotheses a method where only the electrodynamic properties of the body come [6] into account, offer surprising characteristics. According to these equations, the laws that govern the propagation of displacement currents in a dielectric medium are exactly those which obey the infinitely small displacement of a perfectly elastic body; in particular, uniformly moving currents behave absolutely like vibrations of the ether which optics then attributed to the light phenomena.

But there is more. The velocity of the displacement current in a vacuum can be measured by purely electrical experiments; and this speed, thus determined, is numerically equal to the speed of light in a vacuum. Therefore, it is only a simple analogy between uniform displacement flux and luminous vibration which imposes itself on the spirit of the physicist; immediately, he is led to believe that light vibrations

---

[5]G. Kirchhoff, *Ueber die Bewegung der Elektricität in Leitern* (POGGENDORFF'S ANNALEN, Bd. CII, 1857). [English translation: Graneau and Assis (1994)].

do not exist. To periodic displacement currents, he attributes the phenomena that these vibrations were used to explain, often in a less than fortunate way; thus creating the *electromagnetic theory of light*, Maxwell made optics a province of electrodynamics.

Surprising for its consequences, the electrodynamics that Maxwell inaugurated was even more so by the unusual way that it followed its author into science.

Physical theory is a symbolic construction of the human mind intended to give a representation—a synthesis as complete, as simple, and as logical as possible—of the laws that experience has discovered. To each new quality of bodies, it matches a quantity where the various values are used to identify the various states, the various intensities of this quality. Among the different quantities that he considers, he establishes connections using mathematical propositions that seem to translate the simple properties and most essential qualities of which these quantities are the signs; then, deducing from these *hypotheses*, by rigorous reasoning, the consequences that they implicitly contain, he compares these consequences to the laws that the experimenter has uncovered. When a large number of these theoretical consequences represent, in a very approximate way, a large number of experimental laws, the theory is good. [7]

The theory must give of the physical world a description as simple as possible; it must therefore restrict as far as possible the number of properties that it regards as irreducible qualities and that it describes by means of particular quantities, the number of laws it regards as primary and of which it makes hypotheses. It must appeal to a new quantity, accept a new hypothesis, only when inescapable necessity compels it.

When the physicist discovers facts unknown to him, when his experiences have allowed him to formulate laws that the theory did not foresee, he must first search with great care if these laws can be presented, to the required degree of approximation, as consequences of the accepted hypotheses. It is only after having acquired the certainty that the quantities previously handled by the theory can serve as symbols to the observed qualities, that the received hypotheses can result from the established laws, that he is allowed to enrich physics with a new quantity, to the complicating of a new hypothesis.

These principles are the essence itself of our physical theories. If one were to miss it, the difficulty which is often encountered is whether or not a law, discovered by observation, following accepted hypotheses or not, too frequently attached to a laziness of the mind, would lead physicists to look at each new property as an irreducible quality, each new law as a first hypothesis, and our science would soon deserve all the reproaches that contemporaries of Galileo and Descartes addressed to the physics of the School.[6]

The founders of electrodynamics are carefully conformed to these principles. To represent the properties of electrified bodies, it was sufficient for Coulomb and Poisson to make use of a single quantity, electric charge, to impose on electric charges a single hypothesis, Coulomb's law. When Ampère discovered that attractive or repulsive actions are exerted between wires carrying electric currents, physicists

---

[6][Who, for example, explained why a sleeping pill makes one sleep by saying it has the "irreducible quality" of "*vis dormitiva*" or "sleeping power"].

sought first whether they could represent these actions by electrical charges properly distributed on wires and repelling each other according to [8] Coulomb's formula. Ampère attempted this. He did not deny that some facts of experience, and in particular the phenomena of electromagnetic rotation, discovered by Faraday, were proof that he could succeed; then he only wanted that the intensity of the electrical current take place with the electric charge. So, he only proclaimed the laws of electrodynamics were first laws, in the same way as Coulomb's law.

To create the electrodynamics of the dielectric body, Maxwell took a back-step.

At the time when Maxwell introduced in electrodynamics a new quantity, the displacement current, at the moment where he marked, as key hypotheses, the mathematical form of the laws to which this quantity should be submitted, no duly observed phenomenon required the extension of the theory of currents. It was enough to represent, if not all phenomena until then known, at least all those whose experimental study had arrived at a sufficient degree of sharpness. No logical necessity urged Maxwell to imagine a new electrodynamics. For guides, he had only analogies, the desire to provide the work of Faraday with an extension similar to what the work of Coulomb and Poisson received from the electrodynamics of Ampère, and possibly also an instinctive sense of the electrical nature of light. It took many years of research and engineering for Hertz to discover phenomena that reflected his equations, so his theory happened to be a form devoid of any material. With incredible imprudence, Maxwell reversed the natural order according to which theoretical physics evolves; he did not live long enough to see the discoveries of Hertz transform his imprudent boldness in prophetic divination.

Entering into science by an unusual route, Maxwell's electrodynamics does not seem less strange when one follows the developments in the writings of its author. [9]

We note at the outset that the writings of Maxwell describe not a single electrodynamics, but at least three distinct electrodynamics.

The first writing by Maxwell[7] is intended to establish in clear light the analogy between the equations that govern various branches of physics, an analogy which seemed to suggest new inventions. "By a physical analogy I mean that partial similarity between the laws of one science and those of another which makes each of them illustrate the other."[8] The analogy, already noticed by Huygens, between acoustics and optics, contributed greatly to the progress thereof. Maxwell takes as his starting point the theory of heat conductivity, or rather the theory of the motion of a fluid in a resistant medium, a simply changing the notation, which does not alter the form of the equations. From these equations, by way of analogy, Ohm had earlier derived the laws of electric motion in conductive bodies; by a similar process, Maxwell deduced a theory of polarization of dielectric bodies.

---

[7]J. Clerk Maxwell, *On Faraday's Lines of Force*, read at the Philosophical Society of Cambridge on 10 December 1855 and 11 February 1856 (TRANSACTIONS OF THE CAMBRIDGE PHILOSOPHICAL SOCIETY, vol. X, part. I, pp. 27–83.—THE SCIENTIFIC PAPERS OF JAMES CLERK MAXWELL, t. I, pp. 156–219; Cambridge, 1890).

[8][*ibid.* p. 156].

The first memoir of Maxwell intended only to *illustrate* the theory of dielectrics by comparing the equations that govern it with the equations that govern other parts of physics. The second[9] aims to be a *mechanical model* that *describes* or *explains* (for an English physicist, the two words have the same meaning)[10] electric and magnetic action.

We know the constitution that, in this memoir, Maxwell [10] assigns to every body: *cells*—whose very thin walls are formed from a perfectly elastic and incompressible solid and contain a perfect, equally incompressible fluid—that animate rapid vortical movements. These vortical movements represent the magnetic phenomena; at each point, the instantaneous axis of vortical motion marks the direction of magnetization. The live force[11] of the rotational motion of the fluid that fills a volume element is proportional to the magnetic moment of this volume element. As for the elastic solid that forms the walls of the cells, the forces acting upon it distort it in various ways. The *displacements* that the various parts experience represent the *polarization* introduced by Faraday to account for the properties of the dielectric media.

To leave aside any presumption on the mechanical constitution of media where electric and magnetic events occur; to take as a unique starting point for the laws that experience has firmly established and that all physicists accept; to transform, then, by mathematical analysis the consequences of these laws so that the formulas are, so to speak, modeled on the equations to which the hypothesis of cells led; thus, to highlight the absolute equivalence between this mechanical interpretation and the commonly accepted electrical theories; to admit this doctrine to the highest degree of likelihood that can attain such an explanation: this appears to have been the purpose of Maxwell in his later publications concerning electricity. Likewise, it seems to be the main purpose of the large memoir entitled: *A Dynamical Theory of the Electromagnetic Field*[12] and of the *Treatise on Electricity and Magnetism*[13] which, in a certain way, this memoir outlines. [11]

He writes in the preface of the first edition[14]:

> In the following Treatise I propose to describe the most important of these phenomena, to shew how they may be subjected to measurement, and to trace the mathematical connexions of the quantities measured. Having thus obtained the data for a mathematical theory of electromagnetism, and having shewn how this theory may be applied to the calculation of phenomena, I shall endeavour to place in as clear a light as I can the relations between

---

[9]J. Clerk Maxwell, *On Physical Lines of Force* (PHILOSOPHICAL MAGAZINE, 4th series, t. XXI, pp. 161–175, 281 291, 338 à 348; 1861. Tome XXIII, pp. 12–24, 85–95; 1862.—THE SCIENTIFIC PAPERS OF JAMES CLERK MAXWELL, vol. I, pp. 451–513; Cambridge, 1900).

[10]*L'École anglaise et les Théories physiques* (REVUE DES QUESTIONS SCIENTIFIQUES, 2e série, tome II, 1893).

[11][*Force vive* or mv^2, related to the kinetic energy $mv^2/2$].

[12]J. Clerk Maxwell, *A Dynamical Theory of the Electromagnetic Field*, read at the Royal Society of London on 8 December 1864 (PHILOSOPHICAL TRANSACTIONS, vol. CLV, pp. 459 à 512; 1865.— THE SCIENTIFIC PAPERS OF JAMES CLERK MAXWELL, t. I, pp. 526 à 597; Cambridge, 1890).

[13]J. Clerk Maxwell, *Treatise on Electricity and Magnetism*, 1st edition, London, 1873.—2° edition, London, 1881.—Traduit en français par G. Seligmann-Lui, Paris, 1885–1889.

[14][*ibid*. p. v–vi].

the mathematical form of this theory and that of the fundamental science of Dynamics, in order that we may be in some degree prepared to determine the kind of dynamical phenomena among which we are to look for illustrations or explanations of the electromagnetic phenomena.

Comparing mathematical forms, by which the various branches of physics are symbolized; constructing mechanisms to imitate the effects that he seems hard-pressed to reduce to figure and movement; grouping the experimental laws into theories composed in the image of dynamics; as many methods that it is legitimate to follow, provided that he does so with rigor and precision; as long as he desires to put the well-established laws into the form that algebraic analogy or mechanical interpretation provides, he will never cause the alteration or the rejection of a part, however small it may be, of these laws. These methods, moreover, appear to be particularly suitable to illuminate the part of physics to which all three are applied, when their conclusions merge in a harmonious agreement.

This agreement, unfortunately, does not occur in the work of Maxwell. The various theories of the Scottish physicist are irreconcilable with the traditional doctrine; they are irreconcilable with each other. At each time, between the best established, most universally accepted laws of electricity, of magnetism, and the equations that the algebraic analogy or mechanical interpretation imposes, disagreement breaks out, shouting; at each moment, it seems that subsequently even his reasonings and his calculations corner Maxwell in an impossibility, in a contradiction; but at the time when the contradiction will [12] become manifest, when the impossibility will jump out to all eyes, Maxwell made a troublesome term disappear, changed an unacceptable sign, transformed the meaning of a letter; then, the dangerous step passed, the new electric theory, enriched with a fallacy, continued its deductions.

The word of "encouragement" on the subject of demonstrative methods used by Maxwell has been prounounced[15]; we do not wish to subscribe to this judgment. Maxwell's mistakes in logic were, we must believe, unconscious mistakes; but, admittedly, a renowned physicist has never been, more than Maxwell, blindly infatuated with his own hypotheses, increasingly deaf to denials of acquired truths. No one has more completely disregarded the laws governing the rational development of physical theories. "I have therefore taken the part of an advocate rather than that of a judge," the author of the *Treatise on Electricity and Magnetism* wrote.[16] He was, for his dynamical explanation of electrical phenomena, an advocate stubbornly convinced of the right of his client; he strictly disregarded witnesses; he has forgotten that at the time of submitting a hypothesis to the sovereign control of laws that experience has verified, the physicist must be to his own ideas the most impartial and most severe of judges.

In the preface to one of the books[17] that he dedicated to the work of Maxwell, H. Poincaré expressed himself thus:

---

[15]H. Poincaré, COMPTES RENDUS, t. CXVI, p. 1020; 1893.

[16]J. Clerk Maxwell, *Treatise on Electricity and Magnetism*. Preface of the first edition [p. xii].

[17]H. Poincaré, *Électricité et Optique*. I. *Les théories de Maxwell et la théorie électromagnétique de la lumière*, préface, p. v; Paris, 1890.

The first time a French reader opens Maxwell's book, a sense of unease, and often even defiance, merges initially with his admiration. It was only after prolonged business and the price of many efforts that this feeling dissipates. Some eminent minds remain the same forever. [13]

Why do the ideas of the English scholar have so much trouble acclimating here? It is without doubt that the education received by most of the enlightened French are disposed to taste precision and logic before anything else.

The old theories of mathematical physics gave us in this respect complete satisfaction. All of our masters, from Laplace to Cauchy, proceeded in the same way. Starting from clearly stated hypotheses, they deduce the consequences with mathematical rigour and then compare them with experience. They seem to want to give each of the branches of physics the same rigour as celestial mechanics.

For a mind accustomed to admire such models, a theory is hardly satisfactory. Not only will he not tolerate any appearance of contradiction, but he will require that its various parts be logically linked to each other and that the number of hypotheses be reduced to a minimum.

...The English scholar does not seek to build a building unique, definitive, and well ordered edifice. It seems rather that he raises a large number of provisory and independent constructs, between which communications are difficult and sometimes impossible.

...We should therefore not boast to avoid any contradiction; but we must come to terms with it. Two conflicting theories can, indeed—provided they do not mix and that are not seeking the bottom of things—be both useful instruments of research, and perhaps reading Maxwell would be less suggestive if it had not opened both new and divergent pathways.

We are among those who cannot take their side of the contradiction.

Of course—and we agree with the opinion of H. Poincaré on this point—we do not regard theoretical physics as a branch of metaphysics; for us, it is only a schematic representation of reality. Using mathematical symbols, it classifies and directs the laws that experience has revealed; it condenses these laws into a small number of hypotheses; but the knowledge it gives us from the outside world is [14] neither more penetrating nor of a different nature than the knowledge provided by experience.

However, he does not conclude that theoretical physics is beyond the laws of logic. It deserves the name of science on the condition of being *rational*. He is free to choose its hypotheses as he pleases, provided that these hypotheses are not redundant or contradictory; and the chain of deductions that connects to the hypotheses the truths of the experimental order must contain no link of dubious strength.

A single physical theory which, from the smallest possible number of compatible hypotheses between them, would derive, by impeccable reasoning, all known experimental laws is obviously an ideal perfection which the human mind will never reach; but if it cannot reach this limit, it must constantly be directed. If various parts of physics are represented by theories unconnected with each other, or even by theories that contradict each other when they meet in a common domain, the physicist must regard this disparity and contradiction as transitory evils; he must endeavour to substitute unity for the disparate, logical agreement for contradiction; he should never have to take sides.

Without a doubt, one should not ask a genius physicist about the road that has led him to a discovery. Some, of which Gauss is the perfect model, always linked their thoughts in a perfect order and offer to our reason no new truth that they do not support with the most rigorous demonstrations. Others, like Maxwell, proceed by leaps and bounds, and if they deign to support the views of their imagination with some evidence, such evidence is, too often, insecure and outdated. Some others are entitled to our admiration. But if the unforeseen intuitions of the latter surprise us more than the majestically ordered deductions of the first, it would be wrong to see in them, more than they, the mark of a genius. If the Maxwells are more "suggestive" than the Gausses, it is because they have not bothered to complete their inventions; after having affirmed new propositions, they have often left us with the difficult task of transforming them into truth. [15]

We especially should carefully preserve ourselves from an error that is in fashion today in a certain school of physicists. It consists in regarding illogical and inconsistent theories as better working instruments, as the more fecund methods of discovery than logically constructed theories. This error would with difficulty be allowed in the history of science. I do not know whether Maxwell's electrodynamics contributed more to the development of the physics than the electrodynamics than Ampere's electrodynamics, this perfect model of the theories that geniuses from the elevated school of Newton at the beginning of the century built.

When, therefore, we find ourselves in the presence of a theory that offers contradictions, this theory being the work of a man of genius, our task is to analyze and discuss until we manage to distinguish clearly, on the one hand, the propositions likely to be logically demonstrated and, on the other hand, statements that offend logic and which must be transformed or rejected. In pursuing this task of criticism, we must guard against the narrowness of mind and petty corrections which would make us forget the merit of the inventor; but, more importantly, we must guard against this blind superstition which, for admiration of the author, would hide the serious defects of the work. He is not so great a genius that he surpasses the laws of reason.

These are the principles that have guided us in the critique of the work of Maxwell.

Part I
The Electrostatics of Maxwell

Chapter 2
The Fundamental Properties of Dielectrics. The Doctrines of Faraday and Mossotti

2.1 The Theory of Magnetization by Induction, Precursor to the Theory of Dielectrics

The theory of magnetism has influenced to such a point the development of our knowledge regarding dielectric bodies that we must, first of all, say a few words about this theory.

Aepinus represented magnets as bodies on which two magnetic fluids, equal in amount, are separated such that the one fluid is at one end of the bar, the other fluid at the other end. Coulomb[1] changed this way, universally accepted in his time, of seeing things. [18] He said:

> I believe that one could reconcile the result of experiments with the calculations by making a few changes to the hypotheses; here is one that seems to explain all the magnetic phenomena of which the preceding tests give accurate measurements. It consists in assuming, in the system of Aepinus, that the magnetic fluid is withdrawn in each molecule or integral part of the magnet or steel; that fluid can be transported from one end to the other of this molecule, giving each molecule two poles, but this fluid may not move from one molecule to another. Thus, for example, if a magnetic needle were very small in diameter, or if each molecule could be regarded as a small needle whose north end would be united to the south end of the needle that precedes it, then there are only the two ends, n and s, of the needle that would give signs of magnetism; thus it would only be at both ends where one of the poles of the molecules would not be in contact with the opposite pole of another molecule.
>
> If such a needle were cut into two parts after having been magnetized, in a for example, the end a of part na would have the same force as the end s the whole needle had, and the end s of the part sa would also have the same force that the end n of the whole needle had before being cut.

---

[1]Coulomb, *Septième Mémoire sur l'Électricité et le Magnétisme.—Du Magnétisme* (MÉMOIRES DE L'ACADÉMIE DES SCIENCES pour 1789, p. 488.—COLLECTION DE MÉMOIRES RELATIFS A LA PHYSIQUE, publiés par la Société française de Physique, t. I: *Mémoires de Coulomb*).

© Springer International Publishing Switzerland 2015
P.M.M. Duhem, *The Electric Theories of J. Clerk Maxwell*,
Boston Studies in the Philosophy and History of Science 314,
DOI 10.1007/978-3-319-18515-6_2

> This fact is very accurately confirmed by experience; because if a very long, thin needle is cut into two parts after having been magnetized, each part, tested on a balance, is magnetized to saturation, and although it is magnetized again, it will not acquire a larger force.

Poisson read this passage. He said[2]:

> Before the works of Coulomb, one assumed the two transported fluids, in the process of magnetization, traveled to both ends of compass needles and accumulated at their poles; while, following this illustrious physicist, boreal and austral fluid only experience [19] infinitely small displacements and do not escape from the molecule of the magnetized body to which they belong.

The concept of a magnetic element, thus introduced into physics by Coulomb, is the basis on which the theory given by Poisson, the magnetic induction of the soft iron, rests; here, indeed, is how Poisson sets out[3] the basic hypotheses of this theory:

> Consider a body magnetized by induction, of any shape and dimensions, in which the *coercive* force is zero and which we will call A, for brevity.
>
> From the foregoing, we will look at this body as an assemblage *of magnetic elements*, separated from each other by gaps inaccessible to magnetism, and behold, with respect to these elements, the various hypotheses resulting from the discussion in which we have just entered:
>
> 1. The dimensions of the magnetic elements, and those spaces that insulate them, are unaffected and can be treated as infinitely small relatively to the body A.
> 2. The material of this body places no obstacle to the separation of the two *boreal* and *austral* fluids in the interior of the magnetic elements.
> 3. Portions of the two fluids that the magnetization separates in an any element are still very small relative to the *neutral fluid* that contains this element, and this neutral fluid is never exhausted.
> 4. These portions of fluid, so separated, travel to the surface of the magnetic element where they form a layer whose thickness, variable from one point to another, is everywhere very small and can also be considered infinitely small, even compared to the dimensions of the element.

The theory of magnetization founded by Poisson on these hypotheses is far from perfect, more than a key argument, it lacks rigor or sins against exactitude.[4] But these flaws, to which it was possible to remedy, [20] must not make us forget the results of paramount importance that the theorist definitively introduced into science. Let us recall some of these results, of which we will have to make use in what follows:

Let $d\omega$ be a volume element cut out of any magnet. If it is straight and directed in the magnetic axis of this element, carrying a length equal to the ratio of its magnetic moment by its volume, we get a directed quantity which is the *intensity of magnetization* at a point on the element $d\omega$; M is this size and A, B, C are the components.

---

[2]Poisson, *Mémoire sur la théorie du Magnétisme*, lu à l'Académie des Sciences, le 2 février 1824 (MÉMOIRES DE L'ACADÉMIE DES SCIENCES pour les années 1821 et 1822, t. V. p. 250).

[3]Poisson, *loc. cit.*, p. 262.

[4]*Étude historique sur l'aimantation par influence* (ANNALES DE LA FACULTÉ DES SCIENCES DE TOULOUSE, t. II, 1888).

The components X, Y, Z of the *magnetic field*, at a point (x, y, z) outside the magnet, are given by the formulas

$$X = -\frac{\partial V}{\partial x}, \quad Y = -\frac{\partial V}{\partial y}, \quad Z = -\frac{\partial V}{\partial z},$$

V being the *magnetic potential function* of the magnet; this function is defined by the equality:

$$V = \int \left(A_1 \frac{\partial \frac{1}{r}}{\partial x_1} + B_1 \frac{\partial \frac{1}{r}}{\partial y_1} + C_1 \frac{\partial \frac{1}{r}}{\partial z_1} \right) d\omega_1, \tag{2.1}$$

(x_1, y_1, z_1) being a point of the element $d\omega_1$,
A_1, B_1, C_1, the components of magnetization at this point,
r, the distance of two points (x, y, z) and (x_1, y_1, z_1),
and the integration extending over the entire magnet.

This potential function is identical to that which comes from a *fictional distribution* of magnetic fluid, a density distribution, at each point (x, y, z) of the mass of the magnet,

$$\rho = -\left(\frac{\partial A}{\partial x} + \frac{\partial B}{\partial y} + \frac{\partial C}{\partial z} \right), \tag{2.2}$$

and, at each point of the surface of the magnet, where N_i is the normal directed to the inside of the magnet, having surface density

$$\sigma = -[A \cos (N_i, x) + B \cos (N_i, y) + C \cos (N_i, z)]. \tag{2.3}$$

[21] At each point inside the magnet, we have

$$\Delta V = -4\pi\rho = 4\pi \left(\frac{\partial A}{\partial x} + \frac{\partial B}{\partial y} + \frac{\partial C}{\partial z} \right). \tag{2.4}$$

At each point of the surface of the magnet, we have

$$\frac{\partial V}{\partial N_i} + \frac{\partial V}{\partial N_e} = -4\pi\sigma = 4\pi[A \cos (N_i, x) + B \cos (N_i, y) + C \cos (N_i, z)] \tag{2.5}$$

If a perfectly soft body is subjected to the influence of a magnet, it is magnetized so that the components of magnetization at each point (x, y, z) of the magnet are linked by the following equalities to the potential function of both the inducing and the induced magnetization:

$$A = -K\frac{\partial V}{\partial x}, \quad B = -K\frac{\partial V}{\partial y}, \quad C = -K\frac{\partial V}{\partial z}. \tag{2.6}$$

In these equalities, K is a constant amount for a given body at a given temperature; it is called *coefficient of magnetization* of the body.

This starting point is sufficient to put the problem of magnetization by induction on bodies devoid of a coercive force completely into equations.

These various results, we said, remained committed to science; only equalities (2.6) have been changed. To account for various phenomena presented by highly magnetic bodies, such as soft iron, and, in particular, the phenomenon of *saturation*, G. Kirchhoff proposed[5] replacing the coefficient of magnetization K by a *magnetizing function* $f(M)$ which varies not only with nature and the body temperature, but [22] also with intensity M of the magnetization. Equalities (2.6) are then replaced by the equalities

$$A = -f(M)\frac{\partial V}{\partial x}, \quad B = -f(M)\frac{\partial V}{\partial y}, \quad C = -f(M)\frac{\partial V}{\partial z}. \tag{2.7}$$

For weakly magnetic bodies, this magnetizing function is reduced, as Poisson wanted, to a coefficient of magnetization.

One can, as indicated by Émile Mathieu[6] and later, by H. Poincaré,[7] remove the inaccuracies in reasoning which mar the theory of Poisson and avoid the experimental difficulties which militate against it. However, the same hypotheses on which this theory is based have something naive which shocks the habits of contemporary physicists. W. Thomson said[8]:

> [I]n the present state of science, no theory founded on Poisson's hypothesis of "two magnetic fluids" moveable in the "magnetic elements" could be satisfactory, as it is generally admitted that the truth of any such hypothesis is extremely improbable. Hence it is at present desirable that a complete theory of magnetic induction in crystalline or non-crystalline matter should be established independently of any hypothesis of magnetic fluids, and, if possible, upon a purely experimental foundation. With this object, I have endeavoured to detach the hypothesis of magnetic fluids from Poisson's theory, and to substitute elementary principles deducible from it as the foundation of a mathematical theory identical with Poisson's in all substantial [23] conclusions.

---

[5]G. Kirchhoff, *Ueber den inducirten Magnetismus eines unbegrenzten Cylinders von weichem Eisen* (CRELLE'S JOURNAL FÜR REINE UND ANGEWANDTE MATHEMATIK, Bd. XLVIII, p. 348, 1853.—G. KIRCHHOFF'S ABHANDLUNGEN, p. 103, Berlin, 1882).

[6]É. Mathieu, *Théorie du Potentiel et ses applications à l'Électrostatique et Magnétisme*; 2e partie: *Applications* (Paris, 1886).

[7]H. Poincaré, *Électricité et Optique*, I.—*Les théories de Maxwell et la théorie électromagnétique de la lumière*, leçons professées à la Sorbonne pendant le second semestre 1888–1889, p. 44 (Paris, 1890).

[8]W. Thomson, *On the Theory of Magnetic Induction in Crystalline and Non-Crystalline Substances* (PHILOSOPHICAL MAGAZINE, 4th series, vol. I, pp. 177–186, 1851.—PAPERS ON ELECTROSTATICS AND MAGNETISM, art. XXX, Sect. 604; London, 1872).

Instead of imagining a magnet as a cluster of magnetic particles equally charged by austral and boreal fluid, and embedded in a medium impermeable to magnetic fluids, Sir W. Thomson treats this magnet as a continuous body whose properties depend on the value taken at each point, by a certain directed quantity, the intensity of magnetization. The fundamental hypotheses that characterize this quantity in magnets in general and in bodies devoid of coercive force in particular are equivalent to the diverse equations that are generally admitted today; it makes the developments of the theory of magnetism easier and more elegant, and at the same time satisfying more our desire to make physical hypotheses independent of any supposition about the existence or properties of molecules.

It is, in the study of magnetism, a special point that has certainly influenced the theory of dielectrics and, in particular, has contributed to introducing the idea of Faraday that the ether, empty of all ponderable matter, is endowed with dielectric properties. This point is the study of *diamagnetic* bodies.

Faraday acknowledged that a bar of bismuth took on, at each point, a magnetization directed not as the magnetic field, but in the direction opposite of this field; bismuth is *diamagnetic*.

At first, diamagnetism seems scarcely compatible with the theory of magnetism by Poisson; magnetic particles can be magnetized only in the direction of the field. The contradiction disappears assuming a hypothesis by Edmond Becquerel.[9]

According to this hypothesis, all bodies, even bismuth, would be magnetic; but ether, deprived of any other material, would also be magnetic. Under these conditions, the bodies we call magnetic would be more magnetic than [24] ether; the bodies less magnetic than ether would seem diamagnetic.

The impossibility of properly diamagnetic bodies, manifest in the hypothesis of Poisson, is no longer so when it exposes the foundations of the theory of magnetism as suggested by W. Thomson; nothing, it seems, prevents one from assigning a negative value to the magnetizing function in Eq. (2.7), which become mere hypotheses. Also, in many places in his writings on magnetism, W. Thomson does not bother to treat actual diamagnetic bodies.

The contradictions that would lead to the existence of such bodies appear again when comparing the laws of magnetism to the principles of thermodynamics.

These contradictions were seen for the first time by W. Thomson, in the testimony of Tait[10]:

> The commonly received opinion, that a diamagnetic body in a field of magnetic force takes the *opposite* polarity to that produced in a paramagnetic body similarly circumstanced, is thus attacked by Thomson by an application of the principle of energy. Since all paramagnetic bodies require time for the full development of their magnetism, and do not instantly lose it when the magnetising force is removed, we may of course suppose the same to be true for diamagnetic bodies; and it is easy to see that in such a case a homogeneous non-crystalline diamagnetic sphere rotating in a field of magnetic force would, if it always tended to take the opposite distribution of magnetism to that acquired by iron under the same circumstances,

---

[9]Edmond Becquerel, *De l'action du Magnétisme sur tous les corps* (COMPTES RENDUS, t. XXXI, p. 198; 1850.—ANNALES DE CHIMIE ET DE PHYSIQUE, 3ᵉ série t. XXVIII, p. 283, 1850).

[10]Tait, *Sketch of Thermodynamics* [p. 88].

be acted upon by a couple constantly tending to turn it in the same direction round its centre, and would therefore be a source of the perpetual motion.

John Parker,[11] by similar reasoning, has shown that the existence of the diamagnetic body would be inconsistent with the principle of Carnot. [25]

Finally, E. Beltrami[12] and ourselves[13] arrived at the conclusion that if we can find, on a diamagnetic body placed in a given field, a magnetic distribution that satisfies Eq. (2.7), this distribution corresponds to a state of unstable equilibrium. It is therefore impossible to admit the existence of a diamagnetic body properly so-called and necessary for the hypothesis of Edmond Becquerel: the ether is susceptible to being magnetized.

2.2 The Polarization of Dielectrics

If the hypotheses of Coulomb and Poisson on the constitution of magnetic bodies extremely deviate from the principles in favor with physicists today, their sharpness, their simplicity, the ease with which the imagination could grasp them, should be, for theorists of the beginning of the century, one of the most alluring hypotheses of physics. All properties that we represent today by *directed quantities* were then attributed to *polarized molecules*, i.e. with molecules, at both ends, of opposite qualities; one sought for analogues of *magnetic polarization*.

The idea of comparing to iron, under the influence of the magnet, the insulating substances, such as glass, sulfur or shellac, subject to the action of electrified bodies, has no doubt offered itself to the minds of physicists. Already Coulomb, in the passage following what we already cited, the following[14] this: [26]

> The hypothesis that we just made seems very similar to this well-known electrical experience: when one charges a pane of glass covered with two metal planes; however thin the planes are, if one is away from the glass pane, they give very considerable signs of electricity; the surfaces of the glass, after one discharges the electricity of the linings, are themselves steeped in two contrary currents and form a very good electrophorus; this phenomenon is related somewhat to the thickness that one gives to the glass plane; thus the electric fluid, albeit of a different nature on both sides of the glass, penetrates the surface to an infinitely

---

[11]John Parker, *On Diamagnetism and Concentration of Energy* (PHILOSOPHICAL MAGAZINE, 5th, vol. XXVII, p. 403, 1889).

[12]E. Beltrami, *Note fisico-matematiche, lettera al prof. Ernesto Cesàro* (RENDICONTI DEL CIRCOLO MATEMATICO DI PALERMO, t. III, meeting of 10 March 1889).

[13]*Sur l'aimantation par influence* (COMPTES RENDUS, t. CV, p. 798, 1887)—*Sur l'aimantation des corps diamagnétiques* (COMPTES RENDUS, t. CVI, p. 736, 1888).—*Théorie nouvelle de l'aimantation par influence fondée sur la thermodynamique* (ANNALES DE LA FACULTÉ DES SCIENCES DE TOULOUSE, t. II, 1888).—*Sur l'impossibilité des corps diamagnétiques* (TRAVAUX ET MÉMOIRES DES FACULTÉS DE LILLE, mémoire n° 2, 1889).—*Leçons sur l'Électricité et le Magnétisme*, t. II, p. 221, 1892.

[14]Coulomb, *Septième Mémoire sur l'Électricité et le Magnétisme* (MÉMOIRES DE L'ACADÉMIE DES SCIENCES DE PARIS pour 1789, p. 489. COLLECTION DE MÉMOIRES RELATIFS A LA PHYSIQUE, publiés par la Société française de Physique; t. I: *Mémoires de Coulomb*).

small distance, and this pane looks exactly like a magnetised molecule of our needle. And if now one placed on the other a series of panes in such a way that, in the meeting of the panes, the positive side forms the surface of the first pane located several inches away from the negative surface of the last pane, each surface of the extremities, as experience also proves, will produce, at fairly considerable distances, effects as sensitive as our magnetic needles; although the fluid of each surface of the panes on the extremities penetrates these tiles to an infinitesimally small depth and electrical fluids from all surfaces in contact balance each other, since one of the faces is positive, the other negative.

A few years later, Avogadro[15] also admitted that the molecules of a non-conductive body of electricity are polarized under the influence of a charged conductor. In the terms of Mossotti,[16] "Professor Orioli used induction exercised by one molecule on another, or one thin disk of glass on another, to explain the mode of action of the electrical machine." [27]

But it is to Faraday that we owe the first extensive developments on the electrification of insulating bodies.

Faraday was careful to specify the following about the thoughts that led him to imagine his hypotheses about the constitution of the *dielectric bodies*[17]:

In the long-continued course of experimental inquiry in which I have been engaged, this general result has pressed upon me constantly, namely, the necessity of admitting two forces, or two forms or directions of a force…, combined with the impossibility of separating these two forces (or electricities) from each other, either in the phenomena of statical electricity or those of the current. In association with this, the impossibility under any circumstances, as yet, of absolutely charging matter of any kind with one or the other electricity only, dwelt on my mind, and made me wish and search for a clearer view than any that I was acquainted with, of the way in which electrical powers and the particles of matter are related; especially in inductive actions, upon which almost all others appeared to rest.

Two theories have, by way of analogy, guided Faraday in his hypotheses affecting the polarization of the dielectric body: the theory of magnetism and the theory of electrolytic actions.

Everyone knows about the representation, imagined by Grotthuss, of the state in which a current traversing an electrolyte is situated; each molecule is oriented in the direction of the current, the electrically positive atom on the side of the negative electrode and the electrically negative atom on the side of the positive electrode. But Faraday is struck[18] by the resemblance a voltmeter has with a capacitor. Put a plate of ice between two sheets of platinum; charge one of the leaves of positive electricity and other with negative electricity; you will have a dielectric plate capacitor; [28]

---

[15] Avogadro, *Considérations sur l'état dans lequel doit se trouver une couche d'un corps non conducteur de l'électricité lorsqu'elle est interposée entre deux surfaces douées d'électricité de différente espèce* (JOURNAL DE PHYSIQUE, t. LXIII, p. 450, 1806).—*Second Mémoire sur l'Électricité* (JOURNAL DE PHYSIQUE, t. LXV, p. 130, 1807).

[16] Mossotti, *Recherches théoriques sur l'induction électrostatique envisagée d'après les idées de Faraday* (BIBLIOTHÈQUE UNIVERSELLE, Archives, t. VI, p. 193, 1847).

[17] Faraday, *On Induction*, read at the Royal Society of London, 21 December 1837 (PHILOSOPHICAL TRANSACTIONS OF THE ROYAL SOCIETY OF LONDON, 1838, p. 1.—Faraday's EXPERIMENTAL RESEARCHES IN ELECTRICITY, series I, vol. I, n° 1163, p. 361).

[18] Faraday, *loc. cit.* (EXPERIMENTAL RESEARCHES, 1. 1, p. 361).

now melt the ice; the water will be electrolyzed; you will have a voltameter. From where does this difference come? Simply, from the liquid state of water allowing ions to travel on the two electrodes; as to the electric polarization of particles, one must assume it pre-exists their mobility and that it already occurred in the ice.

...as the whole effect in the electrolyte appeared to be an action of the particles thrown into a peculiar or polarized state, I was led to suspect that common induction itself was in all cases an *action of contiguous particles*, and that electrical action at a distance (i.e. ordinary inductive action) never occurred except through the influence of the intervening matter.

How will these contiguous particles influence each other? Faraday repeatedly describes this action.

Induction appears[19] to consist in a certain polarized state of the particles, into which they are thrown by the electrified body sustaining the action, the particles assuming positive and negative points or parts, which are symmetrically arranged with respect to each other and the inducting surfaces or particles.

The theory[20] assumes that all the *particles*, whether of insulating or conducting matter, are as wholes conductors. That not being polar in their normal state, they can become so by the influence of neighbouring charged particles, the polar state being developed at the instant, exactly as in an insulated conducting mass consisting of many particles.

...The particles of an insulating dielectric whilst under induction may be compared to a series of small magnetic needles, or more correctly still to a series of small insulated conductors. If the space round a charged globe were filled with a mixture of an insulating dielectric, as oil of turpentine or [29] air, and small globular conductors, as shot, the latter being at a little distance from each other so as to be insulated, then these would in their condition and action exactly resemble what I consider to be the condition and action of the particles of the insulating dielectric itself. If the globe were charged, these little conductors would all be polar; if the globe were discharged, they would all return to their normal state, to be polarized again upon the recharging of the globe.

It is clear that Faraday imagines the constitution of dielectric bodies in the exact likeness of what Coulomb and Poisson assigned to magnetic bodies; it does not, however, appear that Faraday thought about bringing to his ideas on electric polarization the consequences to which the theory of magnetization by induction led Poisson.

This reconciliation is shown for the first time, in a succinct but clear manner, in one of the early writings of W. Thomson.[21] He said:

It is therefore necessary that there be a very special action in the interior of solid *dielectric* bodies to produce this effect. It is likely that this phenomenon would be explained by giving the body an action similar to that which would occur if there were no action in the insulating dielectric medium and if there were a very large number of small conducting spheres uniformly distributed in the body. Poisson showed that the electric action, in this case, would be

---

[19]Faraday, *loc. cit.* (EXPERIMENTAL RESEARCHES, vol. I, p. 409).

[20]Faraday, *Nature of the Electric Force or Forces*, read at the Royal Society of London, on 21 June 1838 (PHILOSOPHICAL TRANSACTIONS OF THE ROYAL SOCIETY OF LONDON, 1838, pp. 265 à 282.—EXPERIMENTAL RESEARCHES, série XIV, vol. I,. p. 534).

[21]W. Thomson, *Note sur les lois élémentaires de l'électricité statique* (JOURNAL DE LIOUVILLE, t. X, p. 220, 1845.—Reproduced, with some developments, under the title: *On the Elementary Laws of Statical Electricity*, in CAMBRIDGE AND DUBLIN MATHEMATICAL JOURNAL, nov. 1845, and in PAPERS ON ELECTROSTATICS AND MAGNETISM, art. II, Sect. 25).

quite similar to the action of soft iron magnet under the influence of the magnetized bodies. Based on the theorems he gave with respect to this action, it is easily able to show that if the space between A and B is filled with a mixture thus constituted, the surfaces of equilibrium are the same as when there is only an insulating dielectric medium without dielectric power, but the potential in the interior of A will be smaller than in the latter case, in a ratio that it is easy to determine from the data [30] related to the state of the insulating medium. This conclusion seems to be sufficient to explain the facts that Faraday has observed with respect to dielectric media...[22]

Around the same time, the Italian Society of Sciences, in Modena, began to contest the following question:

Taking as a starting point the ideas of Faraday on electrostatic induction, give a physico-mathematical theory of the distribution of electricity on conductors of various shapes.

It suffices for Mossotti[23] to resolve the problem, to make a kind of transposition of the formulas that Poisson had obtained in the study of magnetism; this transposition was then completed by Clausius.[24]

To accept the ideas of Faraday, Mossotti, and Clausius on the constitution of the dielectric body seems as difficult today as to admit the hypotheses of Coulomb and Poisson about the magnetic body; but it is easy to subject to the polarization theory a theory analogous to what W. Thomson did for the theory of magnetization; it is a theory thus stripped of any consideration of the polarized molecules of which H. von Helmholtz made use.[25]

We note the foundations of this theory.

At the beginning of the study of electrostatics, two types of undirected [31] quantities are enough to define the distribution of electricity on a body; these two quantities were the *solid electric density* σ at each point inside the body and the *surface electric density* Σ at each point on the surface of the body. Even the founders of electrostatics took this notion for that one; they regarded the surface of bodies as having a very thin, but not infinitely thin, electrical layer.

---

[22][Translated from the French].

[23]Mossotti, *Discussione analitica sull'influenza che l'azzione di un mezzo dielettrico ha sulla distribuzione dell'eleitricità alla superfizie dei piu corpi elettrici disseminati in esso* (MÉMOIRES DE LA SOCIÉTÉ ITALIENNE DE MODÈNE, t. XXIV, p. 49, 1850).—Extraits du même (BIBLIOTHÈQUE UNIVERSELLE, ARCHIVES, t. VI, p. 357, 1847).—*Recherches théoriques sur l'induction électrostatique envisagée d'après les idées de Faraday* (BIBLIOTHÈQUE UNIVERSELLE, ARCHIVES, t. VI, p. 193; 1847).

[24]R. Clausius, *Sur le changement détat intérieur qui a lieu, pendant la charge, dans la couche isolante d'un carreau de Franklin ou d'une bouteille de Leyde, et sur l'influence de ce changement sur le phénomène de la décharge* (ABHANDLUNGENSAMMLUNG ÜBER DIE MECHANISCHE THÉORIE DER WARME, Bd. II, ZUSATZ ZU ABHANDL. X, 1867.—THÉORIE MÉCANIQUE DE LA CHALEUR, traduite en français par F. Folie, t. II, ADDITION AU MÉMOIRE, X, 1869).

[25]H. Helmholtz, *Ueber die Bewegungsgleichungen der Elektrieitat für ruhende leitende Körper*, §8 (BORCHARDT'S JOURNAL FÜR REINE UND ANGEWANDTE MATHEMATIK, Bd. LXXII, p. 114, 1870.—WISSENSCHAFTLICHE ABHANDLUNGEN, Bd. I, p. 611).

Later, the study of abrupt drops of the potential level in contact with two different conductors led to the introduction of a third directed quantity, irreducible to previous ones: the *moment of a double layer* at each point of the surface of contact of the two conductors.

These three species of quantities no longer suffice to represent fully the distribution of electricity on a system when this system contains poorly conductive bodies; to complete the representation of a similar system, it is necessary to make use of a new quantity, a directed magnitude that is assigned to each point of a dielectric body and called the *intensity of polarization* at this point.

A dielectric body is thus a body in which there is an intensity of polarization at each point, defined in magnitude and direction, as a magnetic body is a body in which there is an intensity of magnetization, defined in magnitude and direction at each point. The hypotheses to which the intensity of polarization are subjected, moreover, are modeled after the basic hypotheses that characterize the intensity of magnetization. A single hypothesis—essential, it is true—is proper to the intensity of polarization. This hypothesis, to which one is necessarily led by how Faraday and his successors have represented the constitution of dielectrics, is as follows:

A dielectric element, with volume $d\omega$, whose intensity of polarization has components A, B, C, exerts on an electric charge, placed at a finite distance, the same action as two equal electric charges, the one having μ, the other having $-\mu$, placed first at a point M of the element $d\omega$, the second at a point M' of the same element, so that the direction $M'M$ is that of the polarization; and so we have the equality

$$M.\overline{MM'} = \left(A^2 + B^2 + C^2\right)^{\frac{1}{2}} d\omega.$$

[32] On the contrary, it is recognized that *a magnetic element is not on an electric charge*.

Before summing up the consequences that can be drawn from these hypotheses, let us insist a moment still on the transformation that the hypotheses made by the founders of electrostatics have undergone.

Four species of quantities—the solid electric density, the surface electric density, the moment of a double layer, the intensity of polarization—are used today to represent the electrical distribution on a system. The founders of electrostatics—Coulomb, Laplace, and Poisson—made use of only one of these quantities, solid electric density; they admitted it willingly in their theories because they succeeded without difficulty to imagine the density as of a certain fluid; they reduced the other three quantities to this one. Instead of regarding the electrical layer that covers a body as lacking thickness and assigning it a surface density, they imagined it as a finite, though very small, thickness in which electricity has a finite, though very large, solid density; two such layers, identical in sign near the electricity which they are formed, placed a small distance from the each other other, replaced our present double layer, without thickness. Finally, instead of conceiving, at each point of a dielectric, an intensity of polarization of set magnitude and direction, they placed a conductive particle covered with an electrical layer which contained as much positive as negative fluid.

Today, we no longer require of physical theories a simple and easy-to-imagine mechanism which explains the phenomena. We look at them as rational and abstract constructs that are intended to symbolize a set of experimental laws; therefore, to *represent* the *qualities* that we are studying, we accept without difficulty in our theories quantities of any nature, provided only that these quantities are clearly defined, regardless of whether or not the imagination seizes the properties served by these quantities. For example, the concepts of intensity of magnetization or intensity of polarization remain inaccessible to the imagination, which captures very well, on the contrary, the magnetic particles of Poisson, the [33] electric corpuscles of Faraday, covered at both ends, by fluid layers of opposite signs. But the concept of intensity of polarization involves a much smaller number of arbitrary hypotheses than the notion of a polarized particle; it is more completely cleared of any hypothesis on the constitution of matter. Substituting continuity for discontinuity, it lends to simpler and more rigorous calculations; we owe it preference.

2.3 Key Propositions of the Theory of Dielectrics

The principles we have analyzed allow the development of a complete theory of the electrical distribution on systems of conductive bodies and dielectric bodies. We briefly indicate, and without any demonstration,[26] the key proposals which we will have to use later.

We imagine two small bodies, placed at the distance r the one from the other and carrying quantities q and q' of electricity; we conceive these two small bodies placed not in *ether*, i.e. in what would contain a container where one would have made the physical vacuum, but in the *absolute vacuum*, i.e. in a medium identical to the space of the geometers, having length, width and depth, but devoid of any physical property, in particular the power to magnetize or polarize. The distinction is important; indeed, we have seen that the existence of diamagnetic bodies would be contradictory if the faculty of magnetizing were not attributed to ether, according to the hypothesis admitted by Edmond Becquerel; and, since Faraday, all physicists agree to assign dielectric polarization to the ether.

By an extension of Coulomb's laws (experience verifies these laws for a body placed in the air, but it is not conceivable for a body placed in the absolute vacuum), we assume that these two small bodies repel with a force

$$F = \varepsilon \frac{qq'}{r^2},\qquad(2.8)$$

ε being some positive constant.

---

[26]The reader may find these demonstrations in our *Leçons sur l'Électricité et le Magnétisme*, t. II, 1892.

[34] Suppose that an ensemble of electrified bodies is placed in space and let

$$V = \sum \frac{q}{r} \qquad (2.9)$$

be their *potential function*. At any one point (x, y, z) outside the charged conductor, or *inside* one of them, an electric charge μ undergoes an action whose components are μX, μY, μZ, and we have

$$X = -\varepsilon \frac{\partial V}{\partial x}, \quad Y = -\varepsilon \frac{\partial V}{\partial y}, \quad Z = -\varepsilon \frac{\partial V}{\partial z}. \qquad (2.10)$$

Now imagine a set of polarized dielectric bodies. Let $d\omega_1$ be a dielectric element, (x_1, y_1, z_1) a point of this element, and A_1, B_1, C_1 the components of polarization at the point (x_1, y_1, z_1).

$$\overline{V}(x, y, z) = \int \left(A_1 \frac{\partial \frac{1}{r}}{\partial x_1} + B_1 \frac{\partial \frac{1}{r}}{\partial y_1} + C_1 \frac{\partial \frac{1}{r}}{\partial z_1} \right) d\omega_1, \qquad (2.11)$$

where the integration extends over the ensemble of polarized dielectrics. This formula defines, at the point (x, y, z), the *potential function* of this set. In formula (2.11), which recalls exactly the expression (2.1) of the magnetic potential function, r is the mutual distance of two points (x, y, z), (x_1, y_1, z_1).

The electrostatic field created by the dielectric at the point (x, y, z) has for components

$$\overline{X} = -\varepsilon \frac{\partial \overline{V}}{\partial x}, \quad \overline{Y} = -\varepsilon \frac{\partial \overline{V}}{\partial y}, \quad \overline{Z} = -\varepsilon \frac{\partial \overline{V}}{\partial z}. \qquad (2.12)$$

The potential function V, defined by equality (2.11), is identical to the electrostatic potential function that formula (2.9), applied to a certain *fictitious electrical distribution*, defines; in this [35] fictitious distribution, each point (x, y, z) inside the polarized dielectric is assigned a solid density

$$e = - \left(\frac{\partial A}{\partial x} + \frac{\partial B}{\partial y} + \frac{\partial C}{\partial z} \right), \qquad (2.13)$$

and every point on the surface of two different polarized bodies, designated by indices 1 and 2, corresponds to a surface density

$$E = -[A_1 \cos(N_1, x) + B_1 \cos(N_1, y) + C_1 \cos(N_1, z)$$
$$+ A_2 \cos(N_2, x) + B_2 \cos(N_2, y) + C_2 \cos(N_2, z)]. \qquad (2.14)$$

If one of the two bodies, body 2 for example, is incapable of dielectric polarization, it is sufficient, in the previous formula, to suppress the terms in A_2, B_2, C_2.

We see that at any point inside a continuous dielectric, we have

$$\Delta \overline{V} = -4\pi e = 4\pi \left(\frac{\partial A}{\partial x} + \frac{\partial B}{\partial y} + \frac{\partial C}{\partial z} \right), \tag{2.15}$$

while at any point on the surface of two dielectrics, we have

$$\frac{\partial \overline{V}}{\partial N_1} + \frac{\partial \overline{V}}{\partial N_2} = -4\pi E \tag{2.16}$$

$$= 4\pi \left[\; A_1 \cos (N_1, x) + B_1 \cos (N_1, y) + C_1 \cos (N_1, z) \right.$$
$$\left. + \; A_2 \cos (N_2, x) + B_2 \cos (N_2, y) + C_2 \cos (N_2, z) \right].$$

Consider a system where all bodies likely to be charged are good conductive bodies, homogeneous and non-decomposable by electrolysis, and where all the bodies likely to be polarized are perfectly soft dielectrics; on such a system, electrical equilibrium will be ensured by the following conditions:

1. In each of the conductive bodies, we have

$$V + \overline{V} = \text{const.} \tag{2.17}$$

2. [36] At each point of a dielectric, we have

$$\begin{cases} A = -\varepsilon F(M) \dfrac{\partial}{\partial x}(V + \overline{V}), \\[2mm] B = -\varepsilon F(M) \dfrac{\partial}{\partial y}(V + \overline{V}), \\[2mm] C = -\varepsilon F(M) \dfrac{\partial}{\partial z}(V + \overline{V}). \end{cases} \tag{2.18}$$

In these formulas,

$$M = \left(A^2 + B^2 + C^2 \right)^{\frac{1}{2}}$$

is the intensity of polarization at the point (x, y, z) and $F(M)$ is an essentially positive function of M; this function depends on the nature of the dielectric at the point (x, y, z); from one point to the other, it varies continuously or intermittently depending on whether the nature and the state of the bodies vary in a continuous or discontinuous manner.

In general, as a first approximation, we are content to replace $F(M)$ by a *coefficient of polarization F*, independent of the intensity M of the polarization; with this approximation, equalities (2.18) become

$$\begin{cases} A = -\varepsilon F \dfrac{\partial}{\partial x}(V + \overline{V}), \\[2mm] B = -\varepsilon F \dfrac{\partial}{\partial y}(V + \overline{V}), \\[2mm] C = -\varepsilon F \dfrac{\partial}{\partial z}(V + \overline{V}). \end{cases} \qquad (2.19)$$

This immediately leads to two relationships that will have, in this study, a great importance.

In the first place, compared to equality (2.13), equalities (2.19) show that we have, at any point of a continuous dielectric medium, the equality

$$\varepsilon \frac{\partial}{\partial x}\left[\frac{\partial(V + \overline{V})}{\partial x}\right] + \varepsilon \frac{\partial}{\partial y}\left[\frac{\partial(V + \overline{V})}{\partial y}\right] + \varepsilon \frac{\partial}{\partial z}\left[\frac{\partial(V + \overline{V})}{\partial z}\right] = e. \qquad (2.20)$$

[37] In the second place, compared to equality (2.14), equalities (2.19) show that at any point on the surface of two different media, we have

$$\varepsilon F_1 \frac{\partial(V + \overline{V})}{\partial N_1} + \varepsilon F_2 \frac{\partial(V + \overline{V})}{\partial N_2} = E. \qquad (2.21)$$

From these equalities we draw some important consequences. In the case where it is applied to a homogeneous dielectric, the formula (2.20) becomes

$$\varepsilon F \Delta(V + \overline{V}) = e.$$

This equality, combined with equalities (2.15) and

$$\Delta V = 0,$$

satisfied at any point where there is no real electricity, gives the equality

$$(1 + 4\pi \varepsilon F)\Delta(V + \overline{V}) = 0,$$

and since F is essentially positive, this equality is, in turn,

$$\Delta(V + \overline{V}) = 0, \qquad (2.22)$$

and

$$e = 0. \qquad (2.23)$$

Hence the following proposition, demonstrated by Poisson in the case of the magnetic induction and transposed by W. Thomson and Mossotti to the case of dielectrics:

*When a dielectric, homogeneous, and perfectly soft body is polarized by induction,
the fictitious electric distribution that would equal the polarization of this body is a
purely superficial distribution.*

Imagine now that dielectric 1 is in contact along an area with a charged body 2, but
incapable of any polarization. To each point on this surface, [38] two electric surface
densities correspond: a *real* density Σ and a *fictitious* density E; with equalities
(2.16) et (2.21), we can attain the well known equality

$$\frac{\partial V}{\partial N_1} + \frac{\partial V}{\partial N_2} = -4\pi \Sigma$$

and also the equality

$$\frac{\partial V}{\partial N_2} + \frac{\partial \overline{V}}{\partial N_2} = -4\pi \Sigma,$$

which derives from the condition (2.17). We thus obtain equality

$$4\pi \varepsilon F_1 \Sigma + (1 + 4\pi \varepsilon F_1)E = 0. \tag{2.24}$$

*On the surface of contact of a conductor and a dielectric, the density of the actual
electrical layer Σ is to the density of the fictitious electrical layer E in a negative
ratio* $\left(-\frac{1+4\pi\varepsilon F}{4\pi\varepsilon F}\right)$, *larger than 1 in absolute value and only dependent on the nature
of the dielectric.*

The formulas and theorems we have just quickly reviewed pertain to placing into
equations the issues raised by the study of dielectrics. Two of these issues will play
a major role in the discussions that will follow; it is important to recall the solution
in a few words.

The first of these problems concerns capacitors.

Imagine an enclosed capacitor. At any point of the internal armature, the sum $(V +
\overline{V})$ has the same value U_1, while at any point of the external armature, it has the value
U_0. The gap between the two armatures is occupied by a homogeneous dielectric D
where F is the coefficient of polarization. It is shown without difficulty that, in these
circumstances, the internal armature becomes covered with a real electric charge Q
given by the formula

$$Q = \frac{1 + 4\pi \varepsilon F}{4\pi} A(U_1 - U_0),$$

A being a quantity that depends only on the geometric shape [39] of the space between
the two armatures. The *capacitance of the capacitor*, i.e. the ratio

$$C = \frac{Q}{\varepsilon(U_1 - U_0)},$$

has the value

$$C = \frac{1 + 4\pi \varepsilon F}{4\pi \varepsilon} A. \tag{2.25}$$

Take a capacitor of identical shape to the previous one and place between the armatures of this capacitor a new dielectric D', having a coefficient of polarization F'; the capacitance of this second capacitor will have the value

$$C' = \frac{1 + 4\pi\varepsilon F'}{4\pi\varepsilon}A.$$

As Cavendish did it, in 1771, in some researches[27] that remained unpublished for one hundred years, so Faraday[28] did it again as early as 1837, experimentally determining the ratio of the capacitance of the second capacitor to the capacitance of the first; the result of this measurement will be the number

$$\frac{C'}{C} = \frac{1 + 4\pi\varepsilon F'}{1 + 4\pi\varepsilon F}. \tag{2.26}$$

This number will only depend on the nature of two dielectrics D and D'; this number is given the name of *specific inductive capacity of the dielectric D', relative to the dielectric D*.

By definition, the *absolute specific inductive capacitance* of a dielectric D is the number $(1 + 4\pi\varepsilon F)$; for a non-polarizable medium, it is equal to 1. [40]

The consideration of the second problem is more strictly needed when one considers ether as susceptible to dielectric polarization.

Electrostatics as a whole is built assuming that conductive or dielectric bodies are isolated in the absolute vacuum. If one accepts the hypothesis that we have just discussed, such electrostatics is a pure abstraction, unable to give a picture of reality; but, by a fortunate circumstance, one can easily transform this electrostatics into another where unlimited space, which was empty in the first, is filled by a homogeneous, incompressible, and polarizable ether.

Let F_0 be the coefficient of polarization of the medium in which the studied bodies are immersed. These bodies are of homogeneous conductors of electricity and perfectly soft dielectric. What will the distribution of electricity on such a system in equilibrium be? What forces will the various bodies of which it consists produce?

The following rule reduces the solution of these questions to classical electrostatics:

Replace the polarizable vacuum for the ether; for each conductive body, leave the total electrical charge it bears in reality; to each dielectric, attribute a coefficient φ of fictitious polarization, equal to the excess of its real coefficient of polarization F over the coefficient of polarization F_0 of the ether:

$$\varphi = F - F_0; \tag{2.27}$$

---

[27] *The electrical Researches of the honourable* Henry Cavendish, F. R. S., written between 1771 and 1781; edited by J. Clerk Maxwell (Cambridge).

[28] Faraday, EXPERIMENTAL RESEARCHES IN ELECTRICITY, series XI, *On Induction*; §5. *On Specific Induction*, On Specific Inductive Capacity. Read at the Royal Society of London, 21 December 1837.

finally, replace the constant e by a fictitious constant

$$\varepsilon' = \frac{\varepsilon}{1 + 4\pi \varepsilon F_0}. \tag{2.28}$$

You will get a fictitious system corresponding to the actual given system.

The electrical distribution on the conductive bodies will be the same in the fictional system as in the actual system.

The ponderomotive actions will be the same in the fictional system as in the actual system.

As for the polarization at each point of one of the dielectric bodies [41] other than the ether, it has the same direction in the fictional system and in the actual system; but, to obtain its value in the second system, the value that it has in the former must be multiplied by $\frac{F}{F - F_0}$.

2.4 The Particular Idea of Faraday

From the ideas of Faraday on the polarization we have extracted so far what is more general, what gave birth to the theory of dielectrics. These general ideas are far from representing, in their fullness, the thought of Faraday. Faraday professed, in addition, a very particular opinion on the relationship that exists between the electric charge comprising a conductor and the polarization of the dielectric medium in which the conductor is immersed. This opinion of Faraday did not escape Mossotti, which he adopted; on the other hand, it seems to have struck no contemporary physicist. Heinrich Hertz[29] has exhibited this opinion, observing that it is a limiting case of the theory of Helmholtz, already reported by the great physicist; but neither Helmholtz, nor Hertz, attributed it to Faraday and Mossotti.

For him who reads Faraday with careful attention, it is clear that he admitted the following law:

When a dielectric medium is polarized under the action of charged conductors, at each point on the surface of contact of a conductor and dielectric, the density of the fictitious surface layer that covers the dielectric is EQUAL AND OPPOSITE IN SIGN *to the density of the actual electrical layer that covers the conductor:*

$$E + \Sigma = 0. \tag{2.29}$$

Faraday wrote to Dr Hare[30]:

---

[29] Heinrich Hertz, *Untersuchungen über die Aushreitung der elektrischen Kraft: Einleitende Uebersicht*; Leipzig, 1892. [English translation: Hertz(1893)]—Traduit en français par M. Raveau (LA LUMIÈRE ÉLECTRIQUE, t. XLIV, pp. 285, 335 et 387; 1892).

[30] Faraday, *An Answer to Dr Hare's Letter on Certain Theoretical Opinions* (SILLIMANN'S JOURNAL, vol. XXXIX, p. 108; 1840.—EXPERIMENTAL RESEARCHES IN ELECTRICITY, vol. II, p. 268; London, 1844).

Using the word charge in its simplest meaning, I think that a body *can* be [42] charged with one electric force without the other, that body being considered in relation to itself only. But I think that such charge cannot exist without induction, or independently of what is called the development of an equal amount of the other electric force, not in itself, but in the neighbouring consecutive particles of the surrounding dielectric, and through them of the facing particles of the uninsulated surrounding conducting bodies, which, under the circumstances, terminate as it were the particular case of induction.

It is the existence, in the immediate vicinity of each other, of these two layers, equal in density and opposite in sign, that the possibility is due, for Faraday, of maintaining an electrical layer at the surface of a conductor.

Since the theory assumed the medium which surrounds conductive bodies to be perfectly insulating, it does not seek what force keeps the electrical layer adhering to the surface of the conductor; what maintains it is the property attributed to the medium for not allowing the passage of electricity. If we can talk about the *pressure* that the medium exerts on the electricity for maintaining it, it is in the sense where we talk about mechanical strength of binding; this pressure is the electromotive action that *should be* applied to the electrical layer so that it remains on the surface of the conductor, *if the medium ceased to be insulated*. This idea seems to have been clearly perceived by Poisson[31]; he said:

The pressure that the fluid exerts against the air that contains it is partly composed of the repulsive force and the thickness of the layer; and since one of these elements is proportional to the other, it follows that pressure changes on the surface of an electrified body and is proportional to the square of the thickness or the amount of electricity accumulated at each point on this surface. The air impermeable to electricity must be regarded as a vessel whose shape is determined by that of the electrified body; the fluid contained in this vessel exerts against the walls different pressures [43] at different points, so the pressure that occurs at certain points is sometimes very big and infinite compared to what others experience. In places where the pressure of the fluid overcomes the resistance of the air that opposes it, the air yields, or, if desired, the tank bursts, and fluid flows through such an opening. It is what happens at the end points and sharp edges of angular bodies.

Faraday does not understand the thought of Poisson; he confuses the resistance that the air opposes to the escape of electricity, in virtue of its non-conductibility, with the *atmospheric pressure*, i.e. with the resistance that this same air opposes to the movement of the material masses, under gravity and inertia; and, easily interpreted as the explanation, he draws advantage for his theory which attributes to the action of the layer spread on the dielectric the equilibrium of the layer covering the conductor. He said[32]:

Here I think my view of induction has a decided advantage over others, especially over that which refers the retention of electricity on the surface of conductors in air to the pressure of the atmosphere. The latter is the view which, being adopted by Poisson and Biot is also, I believe, that generally received; and it associates two such dissimilar things, as the ponderous

---

[31] S. D. Poisson, *Mémoire sur la distribution de l'électricité à la surface des corps conducteurs*, lu à l'Académie des sciences le 9 mai et le 3 août 1812 (MÉMOIRES DE LA CLASSE DES SCIENCES MATHÉMATIQUES ET PHYSIQUES in the year 1811, MÉMOIRES DES SAVANTS ÉTRANGERS, p. 6).

[32] Faraday, EXPERIMENTAL RESEARCHES IN ELECTRICITY, series XII, *On Induction*, vol. I, p. 438.

air and the subtile and even hypothetical fluid. ...Hence a new argument arises[33] proving that it cannot be mere pressure of the atmosphere which prevents or governs discharge, but a specific electric quality or relation of the gaseous medium. It is, hence, a new argument for the theory of molecular inductive action.

Moreover, an attentive reader of *The Experimental Researches in Electricity* easily recognizes, in the hypothesis that we develop [44] at this time, what Faraday intends to articulate when he says that electric action is not exercized at a distance, but only between contiguous particles; he certainly wants to say that no amount of electricity can develop on the surface of a material molecule without a charge of equal and opposite sign developing on the surface facing another extremely close molecule.

Mossotti has also understood the thought of Faraday well. He said[34]:

This physicist, considering the state of molecular electric polarization, thinks that there must be two systems of opposing forces which alternate rapidly and hide alternately in the interior of the dielectric, but that they must manifest two special effects opposed to the ends of the same body. On one side, with the simultaneous action of the two systems of forces that develop in the dielectric body, a force equal and opposite to that with which the same layer tends to expel its atoms is born at each point of the electrical layer that covers the excited body; and the opposition of these two forces makes the fluid that makes up the layer to stay on the surface of the electric body. On the opposite side, where the dielectric body touches or envelopes the surfaces of other surrounding electrical bodies, it exerts a force of a species analogous to that of the electrified body and by means of which these surfaces are brought to the contrary electric state.

Mossotti, having demonstrated the existence of surface layers which are equivalent to a dielectric polarized by induction, adds[35]:

These layers that represent, for the limits of the dielectric body, effects not neutralized by two reciprocal systems of internal forces, exercise, on the surface surrounding the conductive body, actions equivalent to those that these same electrical layers of these same bodies exercise directly between them without the intervention of the dielectric body. This theorem gives us the main conclusion of the question that we proposed. [45] The dielectric body, by means of the polarization of the atmospheres of its molecules, only transmits from one body to the other the action between the conductive bodies, neutralizing the electrical action on one and conveying to the other an action equal to that which the first would have exercised directly.

If it is observed that for Faraday and Mossotti the words *electric action, electric force* are at every moment taken as synonyms of *electric charge* or *electric density*, one cannot recognize, in the passages that we have just quoted, the hypothesis that reflects equality (2.29). So, we can say that this equality expresses the *particular Faraday and Mossotti hypothesis.*

---

[33]Faraday, *ibid.*, p. 445.

[34]Mossotti, *Recherches théoriques sur l'induction électrostatique envisagée d'après les idées de Faraday* (BIBLIOTHÈQUE UNIVERSELLE, ARCHIVES, t. VI, p. 194; 1847).

[35]Mossotti, *Ibid.*, p. 196.

Taken strictly, this hypothesis is not consistent with the principles on which the theory of dielectric polarization is based. We have seen, in effect, as a result of Eq. (2.24), that the density of the actual electrical layer spread on the surface of a conductive body still had a higher absolute value than the density, at the same point, of the fictitious electrical layer which would be equivalent to the polarization of the adjacent dielectric.

But this same equality (2.24) teaches us that the hypothesis of Faraday and Mossotti, unacceptable if taken strictly, can be approximately true; it is what happens if εF_1 is very large compared to $\frac{1}{4\pi}$.

So, we can say that the *hypothesis of Faraday and Mossotti will represent an approximate law if the abstract number εF has, for all dielectrics, an extremely large numeric value.*

Let us examine the consequences to which this hypothesis leads.

The capacitance of a variable capacitor varies little when, in this capacitor, a vacuum is made as perfect as possible; one can therefore admit that the specific inductive capacity of air compared to the ether hardly surpasses unity or that the number $(1 + 4F\pi\varepsilon F)$ relative to the air can be substituted for the number $(1 + 4F\pi\varepsilon F)$ relative to the ether.

Take two electrical charges Q and Q' placed in the ether [46] (practically in the air) and let r be the distance between them; these charges repel with a force which has the value

$$R = \frac{\varepsilon}{1 + 4\pi\varepsilon F_0} \frac{QQ'}{r^2}. \tag{2.30}$$

If one accepts the hypothesis of Faraday and Mossotti, this value differs little from

$$R = \frac{1}{4\pi F_0} \frac{QQ'}{r^2}. \tag{2.31}$$

Suppose that one uses the C. G. S. system of electromagnetic units; that the numbers Q, Q', r—which measure, in this system, the charges and their distances—be numbers of moderate magnitude; and that, for example, they be, all three, equal to 1. Experience shows us that the repulsive force is not measured by a very small number, but, on the contrary, by a large number; the coefficient of polarization F_0 of the ether cannot therefore be regarded as having a very high value in the C. G. S. electromagnetic system. The hypothesis of Faraday then entails the following proposition:

In the C. G. S. electromagnetic system, the constant ε has an extremely large value; each formula can be replaced by the limiting form that one gets when ε is made to grow and surpass any limit.

The experience which we have just discussed tells us, moreover, about the value of F_0. The repulsion of two charges represented by the number 1 in the C. G. S. electromagnetic system, placed at one centimeter of distance the one from the other, is measured approximately by the same number as the square of the speed of light, i.e. the number 9×10^{22}; so, if one accepts the hypothesis of Faraday, we roughly have

$$\frac{1}{4\pi F_0} = 9 \times 10^{22}$$

or

$$F_0 = \frac{1}{36\pi \times 10^{22}}.$$

[47] εF_0 being extremely large compared to $\frac{1}{4\pi}$, we see that, *in the C. G. S. electromagnetic system, ε must be measured by a very large number compared to 10^{22}*.

The specific inductive capacity relative to the ether (practically to the air) of a dielectric is the ratio $\frac{1 + 4\pi \varepsilon F}{1 + 4\pi \varepsilon F_0}$; for all dielectrics known, it has a finite value; it varies between 1 (ether) and 64 (distilled water).

Now, in the theory of Faraday, the specific inductive capacity of a dielectric D'compared to another dielectric D is approximately equal to the ratio between coefficient of polarization F' of the first dielectric and the coefficient of polarization F of the second:

$$\frac{1 + 4\pi \varepsilon F'}{1 + 4\pi \varepsilon F} = \frac{F'}{F}. \tag{2.32}$$

So, for all dielectrics, the ratio $\frac{F}{F_0}$ is understood to be between 1 and 64; in other words, for all dielectrics, the coefficient of polarization F, measured in C. G. S. electromagnetic units, is at most on the order of 10^{-22}.

Helmholtz, having developed a very general electrodynamics, suggested,[36] to find various consequences of Maxwell's theory, an operation that amounts to taking the limit of the equations obtained when εF grows beyond any limit. This supposition, it is seen, immediately reduces to the hypothesis of Faraday and Mossotti. [48]

---

[36]H. Helmholtz, *Ueber die Gesetze der inconstanten elektrischen Ströme in körperlich ausgedehnten Leitern* (VERHANDLUNGEN DES NATURHISTORISCH-MEDICINISCHEN VEREINS ZU HEIDELBERG, 21 January 1870; p. 89.—WISSENSCHAFTLICHE ABHANDLUNGEN, Bd. I, p. 513).—Ueber die Bewegungsgleichungen der Elektricität für ruhende leitende Körper (BORCHARDT'S JOURNAL FÜR REINE UND ANGEWANDTE MATHEMATIK, Bd. LXXII, p. 127 et p. 129.—WISSENSCHAFTLICHE ABHANDLUNGEN, Bd. I, p. 625 et p. 628).—See also: H. Poincaré. *Électricité et Optique*; II. *Les théories de Helmholtz et les expériences de Hertz*, p. vi et p. 103; Paris, 1891.

Chapter 3
The First Electrostatics of Maxwell

3.1 Reminder of the Theory of Heat Conductivity

Before going further and addressing the presentation of the ideas of Maxwell, we pause for a moment to study heat conductivity.

We consider a homogeneous or heterogeneous but isotropic substance.

Let (x, y, z) be a point within this substance.

T, the temperature at this point;

k, the coefficient of heat conductivity at that point.

The flow of heat at this point will have, for the components along the coordinate axes:

$$u = -k\frac{\partial T}{\partial x}, \quad v = -k\frac{\partial T}{\partial y}, \quad w = -k\frac{\partial T}{\partial z}. \tag{3.1}$$

We consider a continuous part of a conductor; an element of volume

$$d\omega = dx\, dy\, dz,$$

carved in this region contains a heat source that emits, in time dt, a quantity of heat $j\, d\omega\, dt$; we can designate j as the *intensity of the source*. We have, according to this definition,

$$\frac{\partial u}{\partial x} + \frac{\partial v}{\partial y} + \frac{\partial w}{\partial z} = j$$

[49] or, in virtue of equalities (3.1),

$$\frac{\partial}{\partial x}\left(k\frac{\partial T}{\partial x}\right) + \frac{\partial}{\partial y}\left(k\frac{\partial T}{\partial y}\right) + \frac{\partial}{\partial z}\left(k\frac{\partial T}{\partial z}\right) + j = 0. \tag{3.2}$$

Now let S be the surface separating two substances, 1 and 2, of different conductivities. The element dS of this surface contains a superficial heat source which in

© Springer International Publishing Switzerland 2015
P.M.M. Duhem, *The Electric Theories of J. Clerk Maxwell*,
Boston Studies in the Philosophy and History of Science 314,
DOI 10.1007/978-3-319-18515-6_3

time dt emits a quantity of heat $J\,dS\,dt$; J is the surface intensity of the source. Then we will have

$$u_1 \cos(N_1, x) + v_1 \cos(N_1, y) + w_1 \cos(N_1, z)$$
$$+ u_2 \cos(N_2, x) + v_2 \cos(N_2, y) + w_2 \cos(N_2, z) = J$$

or, in virtue of equalities (3.1),

$$k\frac{\partial T}{\partial N_1} + k_2\frac{\partial T}{\partial N_2} + J = 0. \tag{3.3}$$

These are the fundamental equations, given by Fourier, that govern the propagation of heat by conduction. We know how the work of G.S. Ohm, later completed by G. Kirchhoff, helped to understand the propagation of electrical current within the conductive bodies. To pass from the first problem to the second, it suffices to replace the heat flow by the flow of electricity, the heat conductivity by electrical conductivity, temperature T by the product εV of the constant of Coulomb's laws and of the electrostatic potential function; finally, to substitute for j and J the ratios $\frac{\partial\sigma}{\partial t}$, $\frac{\partial\Sigma}{\partial t}$, where σ, Σ designate solid and surface electric densities.

A similar extension of the equations of heat conductivity can be used to deal with the diffusion of a salt in an aqueous solution, according to the well known remark of Fick.

An analytical analogy may also be established between certain problems relating to the conductivity of heat and some electrostatics problems.

For example, consider the following problem:

A body C is immersed in a space E. Body C and space E are both homogeneous, isotropic, and conducting, but they have [50] different conductivities: k_2 is the conductivity of the body G, and k is the conductivity of the space E. Body G is assumed to be maintained at a constant temperature, the same in all its points, which we will refer to by A. The various elements of the space E do not contain any other cause of their release or absorption of heat than what comes from their specific heat γ. Each element $d\omega$, of density ρ, thus releases in time dt a quantity of heat $-\rho\,d\omega\,\gamma\frac{T}{t}dt$, ensuring that

$$j = -\rho\gamma\frac{T}{t}.$$

Finally, the state of the medium E is assumed to be stationary. T has, at each point, a value independent of t, which turns the previous equality into

$$j = 0.$$

How, to achieve a similar state, should the sources of heat distribute on the surface of the body G? At various points in space E, what will the value of the temperature T be?

The temperature T, continued throughout the space, shall take, at any point of the body G and the surface which confines it, the constant value A; at any point of space E, it should verify the equation

$$\Delta T = 0,$$

to which (3.2) is reduced, when

$$j = 0$$

and k is assumed to be independent of x, y, z. T being thus determined, Eq. (3.3), which will be reduced to

$$k_2 \frac{T}{N_2} + J = 0,$$

will determine the value of J for every point on the surface that bounds the body C. [51]

This problem is analytically similar to the following one:

A homogeneous and electrified conductor G is immersed in an insulating medium E. What is the distribution of electricity at the surface of this conductor in equilibrium?

To pass from the first issue in the second, it is sufficient to replace, in the solution, the temperature T by the electric potential function V and the quotient $\frac{J}{k_2}$ by the product $4\pi \Sigma$, where Σ designates the surface density of the electric layer that covers the conductor C.

It would be difficult to quote the geometer who first noticed this analogy; the mathematicians at the beginning of the century were so perfectly accustomed to handling differential equations which lead to the various theories of physics that a similar analogy was, so to speak, jumping out at them. In any case, it is stated in some previous works of Chasles[1] and W. Thomson.[2]

3.2 Theory of Dielectric Media, Constructed by Analogy with the Theory of the Conduction of Heat

They sought, in the properties of the dielectric media, a deeper analogy with the laws of heat conductivity.

Having dealt with any problem of conductivity, one would pass to the similar problem of electrostatics by retaining the same equations and by changing the meaning of the letters contained therein according to the following rules:

---

[1] M. Chasles, *Énoncé de deux théorèmes généraux sur l'attraction des corps et la théorie de la chaleur* (COMPTES RENDUS, t. VIII, p. 209; 1839).

[2] W. Thomson, *On the Uniform Motion of Heat in Homogeneous Solid Bodies and its Connexion with the Mathematical Theory of Electricity* (CAMBRIDGE AND DUBLIN MATHEMATICAL JOURNAL, February 1842.—Reprinted in the PHILOSOPHICAL MAGAZINE in 1854 and in the PAPERS ON ELECTROSTATICS AND MAGNETISM, Art. 1).

The temperature T would be replaced by a certain function Ψ; [52] this function Ψ would determine the components P, Q, R of the electrostatic *field* at the point (x, y, z) according to the formulas

$$P = -\frac{\partial \Psi}{\partial x}, \quad Q = -\frac{\partial \Psi}{\partial y}, \quad R = -\frac{\partial \Psi}{\partial z}. \tag{3.4}$$

The coefficient of conductivity k would be replaced by a factor K characterizing the dielectric properties of the medium and that would be called its *specific inductive capacity*.

The components of the flow of heat w, v, w would be replaced by the components f, g, h of a vector that would be called the *polarity* at the point (x, y, z), so that it would be

$$\begin{cases} f = KP = -K\dfrac{\partial \Psi}{\partial x}, \\[2mm] g = KQ = -K\dfrac{\partial \Psi}{\partial y}, \\[2mm] h = KR = -K\dfrac{\partial \Psi}{\partial z}. \end{cases} \tag{3.5}$$

The intensity j of the heat source would be replaced by $4\pi K e$, e being the *solid electric density*, so that Eq. (3.2) would become

$$\frac{\partial}{\partial x}\left(K\frac{\partial \Psi}{\partial x}\right) + \frac{\partial}{\partial y}\left(K\frac{\partial \Psi}{\partial y}\right) + \frac{\partial}{\partial z}\left(K\frac{\partial \Psi}{\partial z}\right) + 4\pi K e = 0. \tag{3.6}$$

In the memoir where he deals with the theory that we now present, Maxwell will never consider the surfaces of discontinuity that separate the various bodies with each other. We can indeed, if you will, suppose that the passage of the various bodies into each other is done in a continuous manner through a very thin layer; physicists have often used this process.

These various rules, if they existed on their own, could be regarded as a simple set of formulas, as purely arbitrary conventions; they lose that character, to take that of an electrostatics, of a physical theory that could be confirmed or contradicted by experience, when joined to the following hypothesis: [53]

The system is the seat of actions that admit for potential the quantity

$$U = \frac{1}{2}\int \Psi e \, d\omega, \tag{3.7}$$

the integral extending over the entire system.

Some connections of this new electrostatics lie in the researches of Faraday. It is, admittedly, not about dielectric bodies, but about the magnetic bodies that he traces; but we know the intimate links between the development of the theory of magnets to the development of the theory of dielectric bodies. Various phenomena, Faraday said,[3]

> led me to the idea that if bodies posses a different degree of *conductive power* for magnetism,
> ...I only state the case hypothetically, and use the phrase *conductive power* as a general
> expression of the capability which bodies may possess of the transmission of magnetic
> force; implying nothing as to how the process of conduction is carried on.

Certain bodies have a greater conductive power than the surrounding medium; these would be magnetic bodies properly so-called. Others would conduct less than the medium; these would be diamagnetic bodies. Faraday also seems to have glimpsed[4] that this theory was not at every point in agreement with the classical theory of the polarization of magnets.

Already, a few years ago, the same ideas of Faraday on electric induction had suggested to W. Thomson[5] some similar insights. He wrote:

> It is, no doubt, possible that such forces at a distance may be discovered to be produced
> entirely by the action of contiguous particles of some intervening medium, and we have an
> analogy for this in the [54] case of heat, where certain effects which follow the same laws
> are undoubtedly propagated from particle to particle.

But if a few vestiges of the idea that we just described can be suspected in the writings of some authors, it is not doubtful that Maxwell has first developed them into a genuine theory; he devoted the first part of his oldest memoir on electricity to this theory.[6]

Maxwell begins by proclaiming the fruitful role of *physical analogy*. He said: "By a physical analogy I mean that partial similarity between the laws of one science and those of another which makes each of them illustrate the other," and he shows how the physical analogy between acoustics and optics has contributed to the progress of the latter science.

He then developed not the theory of the propagation of heat in a medium, but a theory of the motion of a fluid in a durable medium; it only differs from it by the meaning of the letters he employs; but in both, these letters are grouped according to the same formulas.

---

[3]Faraday, *Experimental Researches in Electricity*, XXVIth series, read at the Royal Society of London on 28 Nov. 1850 (EXPERIMENTAL RESEARCHES, vol. III, p. 200).

[4]Faraday, *loc. cit.*, p. 208.

[5]W. Thomson, *On the Elementary Laws of Statical Electricity* (CAMBRIDGE AND DUBLIN MATHEMATICAL JOURNAL. 1845.—PAPERS ON ELECTROSTATICS, Art. II, n° 50 [p. 37]).

[6]J. Clerk Maxwell, *On Faraday's Lines of Force*, read at the Philosophical Society of Cambridge, 10 December 1855 and 11 February 1856 (TRANSACTIONS OF THE CAMBRIDGE PHILOSOPHICAL SOCIETY, vol. X, part, I p. 27; 1864.—SCIENTIFIC PAPERS OF JAMES CLERK MAXWELL, vol. 1, p. 156; Cambridge, 1890).

Maxwell extended these formulations to electricity, in accordance with what we indicate.[7] He says[8]:

> The electrical induction exercised on a body at a distance depends not only on the distribution of electricity in the inductric, and the form and position of the inducteous body, but on the nature of the interposed medium, or dielectric. Faraday expresses this by the conception [55] of one substance having a *greater inductive capacity*, or conducting the lines of inductive action more freely than another. If we suppose that in our analogy of a fluid in a resisting medium the resistance is different in different media, then by making the resistance less we obtain the analogue to a dielectric which more easily conducts Faraday's lines.

3.3 Discussion of the First Electrostatics of Maxwell

When Maxwell, in the explanatory statement that we analyzed, speaks of polarity, electric charge, or potential function, did he intend to deprive these words of the meaning they previously received in electrostatics? Did he mean to define new quantities, essentially distinct from those which bore the same names before him, and intend to replace them in a theory irreducible to the old electrostatics? Many passages of his memoir clearly prove that this is not so; he intends to use the words electric charge, potential function, and polarity in the sense accepted by all. He does not claim to create a new electrostatics, but, by comparison, to illustrate the traditional electrostatics, the theory of polarization of dielectrics such as Faraday and Mossotti have conceived, in imitation of theory of magnetism given by Poisson.

First, speaking of the state of electrostatics at the time when he wrote, Maxwell does not seem to propose altering anything in the accepted formulas; then, he indicates by what change in the meaning of the letters of the formulas we pass from the problem of the movement of a fluid in a resistant medium to the "ordinary" electric problem, an epithet whose employment excludes any intention to revolutionize this branch of physics. Regarding magnets, Maxwell clearly remarks that the two theories in question are, for him, mathematically equivalent. He said[9]:

> A magnet is conceived to be made up of elementary magnetized particles, each of which has its own north and south poles, the action of which upon other north and south poles is governed by laws mathematically identical with those of electricity. Hence the same application of

---

[7]To reconcile our notation with that used by Maxwell in the cited memoir, we need to replace

$$\Psi \quad \text{by} \quad -V,$$
$$e\,d\omega \quad \text{by} \quad dm,$$
$$K \quad \text{by} \quad \frac{1}{K},$$
$$f, g, h \quad \text{by} \quad u, v, w,$$
$$P, Q, R \quad \text{by} \quad X, Y, Z.$$

[8][p. 177].

[9][*ibid.*, p. 178].

[56] the idea of lines of force can be made to this subject, and the same analogy of fluid motion can be employed to illustrate it.

Maxwell develops this analogy and applies it to magnetic bodies considered more conductive than the ambient medium and to diamagnetic bodies regarded as less conductive than this medium, and he adds[10]:

It is evident that we should obtain the same mathematical results if we had supposed that the magnetic force had a power of exciting a polarity in bodies which is in the *same* direction as the lines in paramagnetic bodies, and in the *reverse* direction in diamagnetic bodies.

It is palpable that Maxwell, in relying on an analogy with the equations of heat, simply claimed to give a theory of dielectrics different from the point of view of physical hypotheses but identical with the mathematical equations to the theory that dominates the hypothesis of polarized molecules.

Also, Maxwell does not hesitate to admit[11] that the function Ψ is analytically identical to the electrostatic potential function:

$$\Psi = \int \frac{e}{r}\, d\omega. \tag{3.8}$$

There was only discussion thus far in Maxwell's theory of dielectric bodies; how does Maxwell represent conductive bodies? He said[12]:

If the conduction of the dielectric is perfect or nearly so for the small quantities of electricity with which we have to do,...The dielectric is then considered as a conductor, its surface is a surface of equal potential, and the resultant attraction near the surface itself is perpendicular to it. [57]

Thus, for Maxwell, there is not, strictly speaking, a conductive body; all bodies are dielectrics, which only differ from one to another by the value assigned to K. For the ether of the vacuum, K is equal to 1; for other dielectrics, K is greater than 1; for some, K has a very high value; those are the conductors.

Therefore, the electrostatic problem is as follows:

The function Ψ that defines equality (3.8) must satisfy in all space equality (3.6); once this function Ψ is determined, equalities (3.5) will provide, at each point, the state of the polarization of the medium.

However, equality (3.8), which is a definition, causes the identity

$$\Delta\Psi = 4\pi e,$$

---

[10][*ibid.*, p. 179–80].

[11]J. Clerk Maxwell, Scientific Papers, vol. I, p. 176; Maxwell wrote the equality

$$V = -\sum \frac{dm}{r}$$

which, with his notation, is equivalent to the previous one.

[12][*ibid.*, p. 178].

so that equality (3.6) can also be written

$$\frac{\partial K}{\partial x}\frac{\partial \Psi}{\partial x} + \frac{\partial K}{\partial y}\frac{\partial \Psi}{\partial y} + \frac{\partial K}{\partial z}\frac{\partial \Psi}{\partial z} = 0. \tag{3.9}$$

This condition is *all* that the first electrostatics of Maxwell gives us to determine the function Ψ; however, it is clear that it is insufficient for this purpose. First of all, in a homogeneous medium, where K is independent of x, y, z, it is reduced to an identity and leaves the function Ψ entirely indeterminate in a similar medium. But, even in the event that, to avoid this difficulty, we would reject the existence of any homogeneous medium, or would not take a step to determine Ψ, since if a function Ψ verifies Eq. (3.9), the function $\lambda\Psi$, where λ is a constant, may also satisfy it.

The first electrostatics of Maxwell, thus, has only the appearance of a physical theory; when one follows it closely, it vanishes. [58]

Chapter 4
The Second Electrostatics of Maxwell

4.1 The Hypothesis of Electrical Cells

The first electrostatics was for Maxwell but a mere blueprint; the second electrostatics, which we now explain, is, instead, a developed theory, to which its author returned on several occasions. More closely than the first theory, it is inspired by views of Faraday and especially Mossotti on the constitution of dielectrics.

Faraday considered a dielectric subjected to induction as composed of particles whose two ends carry equal and contrary charges; but he avoided any determined hypothesis on the intrinsic nature of this electricity possessed by the material particles, and by which they can be either polarized or left in the neutral state; he likes to insist on the fact that his theory of induction is independent of any hypothesis about the nature of electricity.

He said[1]:

> My theory of induction makes no assertion as to the nature of electricity, or at all questions any of the theories respecting that subject. It does not even include the origination of the developed or excited state of the power or powers; but taking [59] that as it is given by experiment and observation, it concerns itself only with the arrangement of the force in its communication to a distance in that particular yet very general phenomenon called *static induction*. It is neither the nature nor the amount of the force which it decides upon, but solely its mode of distribution.

Mossotti did not imitate the caution with which Faraday kept away from any hypothesis on the nature of electricity and avoided deciding between the theory which posits two electrical fluids and that which admits a single fluid. A staunch supporter of the ideas of Franklin, he transports them into his exposition of the doctrine of Faraday. He admits that electricity consists of a single fluid, which he calls the *ether*. This fluid exists, to a certain degree of density, in bodies in the

---

[1] M. Faraday, *An Answer to Dr Hare's Letter on Certain Theoretical Opinions* (SILLIMANN'S JOURNAL, vol. XXXIX, p. 108 à 120; 1840.—FARADAY'S EXPERIMENTAL RESEARCHES IN ELECTRICITY, vol. II, p. 262).

© Springer International Publishing Switzerland 2015
P.M.M. Duhem, *The Electric Theories of J. Clerk Maxwell*,
Boston Studies in the Philosophy and History of Science 314,
DOI 10.1007/978-3-319-18515-6_4

neutral state; if it condenses into a region, this region is charged positively; it is charged negatively when the ether is rare; in a dielectric in the neutral state, the ether forms an atmosphere around each of the material particles that cannot leave. When the molecule is subjected to an inductive force, the "ethereal atmosphere[2] condensed at one end exerts a positive and rarefied electrical force at the opposite end, leaving a negative electrical force uncovered."

It is by allowing this passage of Mossotti that Maxwell wrote[3] the following, at the beginning of the presentation of his electrostatics:

Electromotive force acting on a dielectric produces a state of polarization of its parts similar in distribution to the polarity of the particles of iron under the influence of a magnet, and, like the magnetic polarization, capable of being described as a state in which every particle has its poles in opposite conditions. [60]

In a dielectric under induction, we may conceive that the electricity in each molecule is so displaced that one side is rendered positively, and the other negatively electrical, but that the electricity remains entirely connected with the molecule, and does not pass from one molecule to another.

The effect of this action on the whole dielectric mass is to produce a general displacement of the electricity in a certain direction...The amount of the displacement depends on the nature of the body, and on the electromotive force; so that if h is the displacement, R the electromotive force, and E a coefficient depending on the nature of the dielectric,

$$R = -4\pi E^2 h. \tag{4.1a}$$

...These[4] relations are independent of any theory about the internal mechanism of dielectrics...

This passage, where the agreement of theory which will be developed is stated so formally,—on the one hand, with the theory of magnetization by induction given by Coulomb and Poisson, and, on the other hand, with the similar views of Mossotti affecting the polarization of dielectrics—is a piece of information of primary importance on the views of Maxwell. We will find it, in fact, reproduced almost verbatim in all what Maxwell will from now on write regarding electricity, and even in the first chapters of the second edition of his *Treatise*, the last work to which he set his hands.

In the memoir: *On Physical Lines of Force*, which we propose to analyze, Maxwell is not content to accept these results as "independent of any theory." He seeks a combination of fluid bodies and solid bodies that allows him to give a mechanical interpretation; according to the honored word of English physicists, he built a *mechanical model* of dielectrics.

Maxwell admits that any dielectric is a mechanism formed by the means of two substances: an incompressible fluid that lacks [61] viscosity, which he calls *ether*, and a perfectly elastic solid, which he calls *electricity*.

---

[2]Mossotti, *Recherches théoriques sur l'induction électrostatique envisagée d'après les idées de Faraday* (BIBLIOTHÈQUE UNIVERSELLE, ARCHIVES, t. VI, p. 195, 1847).

[3]J. Clerk Maxwell, *On Physical Lines of Force, Part III: The Theory of molecular Vortices applied to statical Electricity* (PHILOSOPHICAL MAGAZINE, January and February 1862.—SCIENTIFIC PAPERS, vol. I, p. 491).

[4]The sign −, in the second member of Eq. (4.1a), comes, as we shall see later, from a clerical error.

Electricity forms very thin cell walls and fills the ether. The ether is animated, within each cell, with vertical movements that explain the magnetic properties of the medium.

> When the electric particles are urged in any direction, they will, by their tangential action on the elastic substance of the cells, distort each cell, and call into play an equal and opposite force arising from the elasticity of the cells. When the force is removed, the cells will recover their form, and the electricity will return to its former position.[5]

In this depiction of the dielectric polarization, the *displacement* of the elastic substance named *electricity* will play exactly the same role as the *displacement of the ethereal fluid* of which Mossotti spoke; at each point, it will measure the *intensity of polarization*.

The elastic cell walls are deformed by the forces that act on them. Let P, Q, R be the components of the force at a point and f, g, h the components of the displacement at the same point; the components f, g, h of displacement depend on the components P, Q, R of the force. How do they depend?

The answer to this question depends on a problem of elasticity which would be very complicated if the shape of the cells were given, and which cannot even be put into equations as long as this form remains unknown; lacking an exact solution, Maxwell was content with a rough approximate solution. He studied the deformation of a single, spherically-shaped cell subjected to a force that is parallel to OZ and has at all points the same value R. It is then that we have

$$R = 4\pi E^2 h, \tag{4.1b}$$

E^2 being a quantity which depends on both of the coefficients of elasticity of the material forming the cells.

Generalizing this result, he admits that we have, in all circumstances, the equalities

$$P = 4\pi E^2 f, \quad Q = 4\pi E^2 y, \quad R = 4\pi E^2 h. \tag{4.2a}$$

[62] In reality, these formulas are not those given by Maxwell, but those that would have provided a correct calculation. As a result of a manifest sign error,[6] he substitutes for these formulas the incorrect formulas

---

[5][*ibid.*, p. 492].

[6]J. Clerk Maxwell, Scientific Papers, vol. I, p. 495. From equations

$$R = -2\pi ma(e + 2f), \tag{100}$$

$$h = \frac{ae}{2\pi}, \tag{103}$$

Maxwell derives the equation

$$R = 4\pi^2 m \frac{e + 2f}{e} h. \tag{104}$$

Moreover, this whole memoir of Maxwell is literally riddled with sign errors.

$$R = -4\pi E^2 h, \tag{4.1a}$$

$$P = -4\pi E^2 f, \quad Q = -4\pi E^2 y, \quad R = -4\pi E^2 h. \tag{4.2b}$$

The formulas that we just wrote are general; they take a particular form when electrical equilibrium is established on the system. In this case, indeed, the electrodynamic theories developed by Maxwell in the memoir that we analyze[7] show that there is a certain function $\Psi(x, y, z)$, such that we have

$$P = -\frac{\partial \Psi}{\partial x}, \quad Q = -\frac{\partial \Psi}{\partial y}, \quad R = -\frac{\partial \Psi}{\partial z}. \tag{4.3}$$

Moreover, if the reasonings of Maxwell demonstrate the existence of this function, they do not inform us in any way of its nature, although Maxwell insinuates the following: "The physical interpretation of Ψ is that it represents the *electric potential* at each point of space."[8] [63]

4.2 The Preceding Principles in the Later Writings of Maxwell

Before following the consequences of these principles and analyzing them further, we will indicate in what form they are found in the writings published by Maxwell after his memoir: *On Physical Lines of Force*.

In 1864, Maxwell published a new, very extensive memoir[9] on electromagnetic actions; there, he himself defined, in the following matter, the spirit which directed the composition of this work. He said[10]:

> I have on a former occasion attempted to describe a particular kind of motion and a particular kind of strain, so arranged as to account for the phenomena. In the present paper I avoid any hypothesis of this kind; and in using such words as electric momentum and electric elasticity in reference to the known phenomena of the induction of currents and the polarization of dielectrics, I wish merely to direct the mind of the reader to mechanical phenomena which will assist him in understanding the electrical ones. All such phrases in the present paper are to be considered as illustrative, not as explanatory.

Without making any hypotheses about the nature of the electrical phenomena, to give to the laws that govern them analogous forms in all respects to cells that affect

---

[7]J. Clerk Maxwell, Scientific Papers, vol. I, p. 482.

[8][*ibid.*].

[9]J. Clerk Maxwell, *A Dynamical Theory of the Electromagnetic Field*, read at the Royal Society of London on 8 December 1854 (Philosophical Transactions, vol. CLV.—Scientific Papers, vol. I, p. 526).

[10]J. Clerk Maxwell, Scientific Papers, vol. I, p. 563.

the equations of dynamics, will precisely be the object of the *Treatise on Electricity and Magnetism*, of which the memoir: *A Dynamical Theory of the Electromagnetic Field* is the draft.

Maxwell shows himself no less respectful of the traditional hypotheses regarding the polarization of dielectrics than in his previous memoir: *On Physical Lines of Force*. He writes,[11] citing Faraday and Mossotti:

> …when electromotive force [64] acts on a dielectric it produces a state of polarization of its parts similar in distribution to the polarity of the parts of a mass of iron under the influence of a magnet, and like the magnetic polarization, capable of being described as a state in which every particle has its opposite poles in opposite conditions.
>
> In a dielectric under the action of electromotive force, we may conceive that the electricity in each molecule is so displaced that one side is rendered positively and the other negatively electrical, but that the electricity remains entirely connected with the molecule, and does not pass from one molecule to another. The effect of this action on the whole dielectric mass is to produce a general displacement of electricity in a certain direction…In the interior of the dielectric there is no indication of electrification, because the electrification of the surface of any molecule is neutralized by the opposite electrification of the surface of the molecules in contact with it; but at the bounding surface of the dielectric, where the electrification is not neutralized, we find the phenomena which indicate positive or negative electrification.
>
> The relation between the electromotive force and the amount of electric displacement it produces depends on the nature of the dielectric, the same electromotive force producing generally a greater electric displacement in solid dielectrics, such as glass or sulphur, than in air.

If one denotes by K the ratio between electromotive force and displacement, there

$$P = Kf, \quad Q = Kg, \quad R = Kh. \tag{4.4}$$

Moreover, in the case where equilibrium is established for the system, the components P, Q, R of the electromotive force are given by the formulas

$$P = -\frac{\partial \Psi}{\partial x}, \quad Q = -\frac{\partial \Psi}{\partial y}, \quad R = -\frac{\partial \Psi}{\partial z}, \tag{4.3}$$

where Ψ is a function of x, y, z, the analytic form of which the electrodynamic reasonings of Maxwell [65] tell us nothing. Maxwell said[12]:

> Ψ is a function of x, y, z, and t, which is indeterminate as far as regards the solution of the above equations, because the terms depending on it will disappear on integrating round the circuit. The quantity Ψ can always, however, be determined in any particular case when we know the actual conditions of the question. The physical interpretation of Ψ is that it represents the *electric potential* at each point of space.

This passage differs little from the one Maxwell wrote about the quantity Ψ, in his memoir: *On Physical Lines of Force*, but by the substitution of the words *electric potential* for the words *electrical voltage*. But, despite the greater accuracy of the

---

[11] J. Clerk Maxwell, IBID., vol. I, p. 531.
[12] J. Clerk Maxwell, SCIENTIFIC PAPERS, vol. I, p. 558.

new term, nothing in the reasonings of Maxwell justifies the analytical identification of the function Ψ with the *electrostatic potential function* of Green; nothing more, not one line of text, nor an equation, shows that Maxwell admitted this assimilation, which is incompatible with many of the results that he reached.

The equations that we just wrote are obviously consistent with those that we have borrowed from the memoir: *On Physical Lines of Force*; they differ only by the substitution of the coefficient K for the product $4\pi E^2$. In addition, the sign error which affected Eqs. (4.1a) and (4.2b) is corrected in Eq. (4.4).

4.3 The Equation of Free Electricity

By the letter e, Maxwell represents, in his memoir: *A Dynamical Theory of the Electromagnetic Field*,[13] "the quantity of free positive electricity contained in unit of volume at any part of the field, then, since this arises from the electrification of the different parts of the field not neutralizing each other."

Paralleling the passage on the dielectric polarization which we have, for the previous section, borrowed from the same memoir, this definition leaves no doubt about the meaning which Maxwell [66] attributes to the letter e; it is the solid density of the fictitious electrical distribution which is equivalent to the dielectric polarization; it is therefore the same as what in Chap. 2 we designated by the letter e.

Secondly, as the displacement (f, g, h) is surely, for Maxwell, the exact equivalent of the intensity of polarization between the components of the displacement and the quantity e, he does not hesitate to write[14] the relationship that Poisson had established between the components of magnetization and the fictitious magnetic density, and which Mossotti had extended to dielectrics:

$$e + \frac{\partial f}{\partial x} + \frac{\partial g}{\partial y} + \frac{\partial h}{\partial z} = 0. \tag{4.5}$$

The former equation is completed by setting the density of *free electricity* in the surface of separation of two dielectrics 1 and 2. In both memoirs that we now analyze, Maxwell never speaks of surfaces of discontinuity; he therefore does not write this equation; but the form is forced, once one accepts, on the one hand, the previous equation and, on the other hand, the equivalence between a surface of discontinuity and a layer of very thin passage. One can added to the previous equation the relationship

$$E + f_1 \cos (N_1, x) + g_1 \cos (N_1, y) + h_1 \cos (N_1, z)$$
$$+ f_2 \cos (N_2, x) + g_2 \cos (N_2, y) + h_2 \cos (N_2, z) = 0. \tag{4.6}$$

---

[13] *Ibid.*, vol. 1, p. 561.

[14] J. Clerk Maxwell, SCIENTIFIC PAPERS, vol. I, p. 561, equality (G).

With Eq. (4.4), Eq. (4.6) becomes

$$E + \frac{1}{K_1}[P_1 \cos(N_1, x) + Q_1 \cos(N_1, y) + R_1 \cos(N_1, z)]$$
$$+ \frac{1}{K_2}[P_2 \cos(N_2, x) + Q_2 \cos(N_2, y) + R_2 \cos(N_2, z)] = 0, \qquad (4.7)$$

while Eq. (4.5) becomes

$$e + \frac{\partial}{\partial x}\frac{P}{K} + \frac{\partial}{\partial y}\frac{Q}{K} + \frac{\partial}{\partial z}\frac{R}{K} = 0, \qquad (4.8)$$

[67] and, in the case where the medium is homogeneous,

$$e + \frac{1}{K}\left(\frac{\partial P}{\partial x} + \frac{\partial Q}{\partial y} + \frac{\partial R}{\partial z}\right) = 0. \qquad (4.9)$$

Maxwell did not write this equation in his memoir: *A Dynamical Theory of the Electromagnetic Field*, but it immediately results from Eqs. (4.4) and (4.5) that he did write.

In the memoir: *On Physical Lines of Force*, he obtains it by different considerations, hardly different from the previous ones, which we need to relate.

The part[15] of this principle [is] that "a variation of displacement is equivalent to a current" such that $\frac{\partial f}{\partial t}, \frac{\partial g}{\partial t}, \frac{\partial h}{\partial t}$ are the components of a current, the *displacement current*, which should be respectively added to the components of the conduction current to form components p, q, r of the total current.

If e be the quantity of free electricity in unit of volume, then the equation of continuity will be

$$\frac{\partial p}{\partial x} + \frac{\partial q}{\partial y} + \frac{\partial r}{\partial z} + \frac{\partial e}{\partial t} = 0.$$

But, by the considerations we will encounter when we study the electrodynamics of Maxwell, he assigns to the components of the conduction current the form

$$-\frac{1}{4\pi}\left(\frac{\partial \gamma}{\partial y} - \frac{\partial \beta}{\partial z}\right), \quad -\frac{1}{4\pi}\left(\frac{\partial \alpha}{\partial z} - \frac{\partial \gamma}{\partial x}\right), \quad -\frac{1}{4\pi}\left(\frac{\partial \beta}{\partial x} - \frac{\partial \gamma}{\partial y}\right),$$

where α, β, γ are three functions of x, y, z. It follows that the previous equation remains exact if we substitute for p, q, r the only components of the displacement current, and that it can be written

$$\frac{\partial}{\partial t}\left(\frac{\partial f}{\partial x} + \frac{\partial g}{\partial y} + \frac{\partial h}{\partial z}\right) + \frac{\partial e}{\partial t} = 0$$

---

[15] J. Clerk Maxwell, SCIENTIFIC PAPERS, vol. I, p. 496.

[68] or, in virtue of equalities (4.2a),

$$\frac{\partial}{\partial t}\left[\frac{\partial}{\partial x}\frac{P}{4\pi E^2} + \frac{\partial}{\partial y}\frac{Q}{4\pi E^2} + \frac{\partial}{\partial z}\frac{R}{4\pi E^2}\right] + \frac{\partial e}{\partial t} = 0, \qquad (4.10)$$

and, in the case of a homogeneous medium,

$$\frac{1}{4\pi E^2}\frac{\partial}{\partial t}\left(\frac{\partial P}{\partial x} + \frac{\partial Q}{\partial y} + \frac{\partial R}{\partial z}\right) + \frac{\partial e}{\partial t} = 0. \qquad (4.11)$$

Until this point of reasoning, one could doubt whether by e Maxwell simply means the density of the fictitious electrical distribution equivalent to the dielectric polarization, or if it includes some real electrification communicated in the medium; a phrase resolves the issue. He said[16]:

$$e = 0 \text{ when there are no electromotive forces.} \qquad (4.1)$$

It is therefore clear that e has the same meaning as in the memoir: *A Dynamical Theory of the Electromagnetic Field*. In addition, from Eqs. (4.11) and (4.11), it is permissible to derive the equations

$$\frac{\partial}{\partial x}\frac{P}{4\pi E^2} + \frac{\partial}{\partial y}\frac{Q}{4\pi E^2} + \frac{\partial}{\partial z}\frac{R}{4\pi E^2} + e = 0, \qquad (4.12)$$

$$\frac{1}{4\pi E^2}\left(\frac{\partial P}{\partial x} + \frac{\partial Q}{\partial y} + \frac{\partial R}{\partial z}\right) + e = 0, \qquad (4.13a)$$

nearly identical in notation to Eqs. (4.8) and (4.9).

We indicate in passing that instead of writing Eq. (4.13a), Maxwell, as a result of the sign error that affects equalities (4.2b), wrote[17]

$$\frac{1}{4\pi E^2}\left(\frac{\partial P}{\partial x} + \frac{\partial Q}{\partial y}\frac{\partial R}{\partial z}\right) = e. \qquad (4.13b)$$

[69]

4.4 The Second Electrostatics of Maxwell is Illusory

The various equalities that we just wrote are general; in the case where equilibrium is established for the system, P, Q, R are related to the function Ψ by equalities (4.3), which give

---

[16]J. Clerk Maxwell, SCIENTIFIC PAPERS, vol. I, p. 497.
[17]*Ibid.*, vol. I, p. 497, equality (115).

$$\begin{cases} f = -\frac{1}{4\pi E^2}\frac{\partial\Psi}{\partial x}, \ g = -\frac{1}{4\pi E^2}\frac{\partial\Psi}{\partial y}, \ h = -\frac{1}{4\pi E^2}\frac{\partial\Psi}{\partial z}, \\ \text{or} \\ f = -\frac{1}{K}\frac{\partial\Psi}{\partial x}, \quad g = -\frac{1}{K}\frac{\partial\Psi}{\partial y}, \quad h = -\frac{1}{K}\frac{\partial\Psi}{\partial z}. \end{cases} \tag{4.14}$$

For equalities (4.3), equalities (4.12) and (4.8) become

$$\begin{cases} \frac{\partial}{\partial x}\left(\frac{1}{4\pi E^2}\frac{\partial\Psi}{\partial x}\right) + \frac{\partial}{\partial y}\left(\frac{1}{4\pi E^2}\frac{\partial\Psi}{\partial y}\right) + \frac{\partial}{\partial z}\left(\frac{1}{4\pi E^2}\frac{\partial\Psi}{\partial z}\right) - e = 0 \\ \frac{\partial}{\partial x}\left(\frac{1}{K}\frac{\partial\Psi}{\partial x}\right) + \frac{\partial}{\partial y}\left(\frac{1}{K}\frac{\partial\Psi}{\partial y}\right) + \frac{\partial}{\partial z}\left(\frac{1}{K}\frac{\partial\Psi}{\partial z}\right) - e = 0. \end{cases} \tag{4.15}$$

Equalities (4.13a) and (4.9) become

$$\frac{1}{4\pi E^2}\Delta\Psi - e = 0, \quad \frac{1}{K}\Delta\Psi - e = 0. \tag{4.16a}$$

Finally, equality (4.7) becomes the second of the equalities

$$\begin{cases} \frac{1}{4\pi E_1^2}\frac{\partial\Psi}{\partial N_1} + \frac{1}{4\pi E_2^2}\frac{\partial\Psi}{\partial N_2} - E = 0, \\ \frac{1}{K_1}\frac{\partial\Psi}{\partial N_1} + \frac{1}{K_2}\frac{\partial\Psi}{\partial N_2} - E = 0. \end{cases} \tag{4.17}$$

If the function Ψ were known, relations (4.14) would determine the components of displacement at each point of the dielectric medium. But how will the function Ψ be determined? By themselves, equalities (4.15), (4.16a) and (4.17) teach us nothing more about this function than equalities (4.14), from which they [70] result. It would be otherwise if some theory, independent from that which provides us Eqs. (4.14), allowed us to express e, E using partial differentials of Ψ, by the relations irreducible to relations (4.15), (4.16a), and (4.17); then, by eliminating e, E among relations (4.15), (4.16a), and (4.17) and these new relations, one would obtain the conditions under which the partial derivatives of the function Ψ would be subject, either at any point of the dielectric medium, or on the surface of separation of the two different dielectrics.

It is by this method that the theory of magnetic induction given by Poisson is developed, the theory of the dielectric polarization conceived in imitation of the previous one by Mossotti.

When in this last theory, we posed the equations of polarization in the form [Chap. 2, equalities (2.19)]

$$A = -\varepsilon F\frac{\partial}{\partial x}(V + \overline{V}),$$

$$B = -\varepsilon F \frac{\partial}{\partial y}(V + \overline{V}),$$

$$C = -\varepsilon F \frac{\partial}{\partial z}(V + \overline{V}),$$

when we derived, for any point of a continuous medium, the relationship [Chap. 2, equality (2.20)]

$$\varepsilon \frac{\partial}{\partial x}\left[F\frac{\partial(V + \overline{V})}{\partial x}\right] + \varepsilon \frac{\partial}{\partial y}\left[F\frac{\partial(V + \overline{V})}{\partial y}\right] + \varepsilon \frac{\partial}{\partial z}\left[F\frac{\partial(V + \overline{V})}{\partial z}\right] - e = 0,$$

$$(4.18)$$

analogous to our equalities (4.15), and, on the surface of separation of two dielectric media, the relation [Chap. 2, equality (2.21)]

$$\varepsilon F_1 \frac{\partial(V + \overline{V})}{\partial N_1} + \varepsilon F_2 \frac{\partial(V + \overline{V})}{\partial N_2} - E = 0, \tag{4.19}$$

analogous to our relations (4.17). But this does not end the solution. The function $(V + \overline{V})$ contained in these formulas is not simply [71] a uniform and continuous function of x, y, z; it is a function whose analytical expression is given in a very precise manner when the electric distribution is given, real or fictitious; and from this analytical expression, in virtue of the theorems of Poisson, two previous independent relationships result. The one [Chap. 2, equality (2.15)], satisfied at any point of a polarized but not electrified continuous dielectric,

$$\Delta(V + \overline{V}) = -4\pi e; \tag{4.20}$$

the other [Chap. 2, equality (2.16)], satisfied on the surface of separation of two such dielectrics,

$$\frac{\partial(V + \overline{V})}{\partial N_1} + \frac{\partial(V + \overline{V})}{\partial N_2} = -4\pi E. \tag{4.21}$$

If then we compare, on the one hand, equalities (4.18) and (4.20), and, on the other hand, equalities (4.19) and (4.21), we find that the partial derivatives of the function $(V + \overline{V})$ must satisfy, at any point of a continuous dielectric, the relation

$$\frac{\partial}{\partial x}\left[(1 + 4\pi e F)\frac{\partial(V + \overline{V})}{\partial x}\right] + \frac{\partial}{\partial y}\left[(1 + 4\pi e F)\frac{\partial(V + \overline{V})}{\partial y}\right]$$
$$+ \frac{\partial}{\partial z}\left[(1 + 4\pi e F)\frac{\partial(V + \overline{V})}{\partial z}\right] = 0$$

and, at the surface of separation of the two dielectric media, the relation

$$(1 + 4\pi\varepsilon F_1)\frac{\partial(V + \overline{V})}{\partial N_1} + (1 + 4\pi\varepsilon F_2)\frac{\partial(V + \overline{V})}{\partial N_2} = 0.$$

It is precisely these partial differential equations which will be used to determine the function $(V + \overline{V})$ and, as a result, the state of polarization of dielectrics.

The same circumstances occur in all analogous problems that mathematical physics provides. Take, for example, the problem of the conductivity of the [72] heat in an isotropic medium. Arising from the hypotheses of Fourier, by designating the solid or surface intensity of the sources of heat by j, J, equation [Chap. 3, Eq. (3.2)]

$$\frac{\partial}{\partial x}\left(k\frac{\partial T}{\partial x}\right) + \frac{\partial}{\partial y}\left(k\frac{\partial T}{\partial y}\right) + \frac{\partial}{\partial z}\left(k\frac{\partial T}{\partial z}\right) + j = 0,$$

satisfied at any point of a continuous medium, and the equation [Chap. 3, Eq. (3.3)]

$$k_1\frac{\partial T}{\partial N_1} + k_2\frac{\partial T}{\partial N_2} + J = 0,$$

satisfied on the surface of separation of two media, results.

The problem of determining the distribution of heat on the system is not put into equations as long as new hypotheses have not connected the intensities j, J at temperature T. Further, we must assume, for example, that the medium does not contain other sources of heat or cold than its own heat capacity, which will in turn be written

$$j = -\rho\gamma\frac{\partial T}{\partial t}, \quad J = 0,$$

ρ being the density of the body and γ its specific heat. The previous equations become, then, for the function T, the partial differential equations

$$\frac{\partial}{\partial x}\left(k\frac{\partial T}{\partial x}\right) + \frac{\partial}{\partial y}\left(k\frac{\partial T}{\partial y}\right) + \frac{\partial}{\partial z}\left(k\frac{\partial T}{\partial z}\right) + j = 0,$$

$$k_1\frac{\partial T}{\partial N_1} + k_2\frac{\partial T}{\partial N_2} = 0,$$

which will be used to determine the distribution of temperature on the system.

There is nothing analogous in the electrostatics of Maxwell. Of the function Ψ appearing in Eqs. (4.15), (4.16a) and (4.17), he knows nothing [73] apart from these equations, if it is not uniform and continuous. He does not have the right to write, regarding this function, any equality that is not a consequence of those which are already given, and, indeed, he writes only what he claims to derive from those. He has therefore no way to eliminate e, E and get an equation that can be used to determine the function Ψ.

It must therefore be recognized that *the second electrostatics of Maxwell does not even put into equations the problem of the polarization of a given dielectric medium.*

4.5 Determination of the Electrostatic Energy

Nevertheless, Maxwell strives to draw a few conclusions from this incompletely posed problem; that is, it must be admitted, in this essay on the constitution of a capacitor, that his imagination, logically carefree, is given the freest career.

The first problem he treats is the formation of *electrostatic energy* or potential actions that are produced in a polarized dielectric.

In his memoir: *On Physical Lines of Force*, Maxwell admits purely and simply[18] that this energy has a value

$$U = \frac{1}{2} \int (Pf + Qg + Rh)d\omega. \tag{4.22}$$

Then, invoking the formulas (4.2b) and (4.3), he finds that U may put in the form

$$U = \frac{1}{2} \int \frac{1}{4\pi E^2} \left[\left(\frac{\partial \Psi}{\partial x}\right)^2 + \left(\frac{\partial \Psi}{\partial y}\right)^2 + \left(\frac{\partial \Psi}{\partial z}\right)^2 \right] d\omega. \tag{4.23a}$$

Formulas (4.2b) are affected by a sign error; if we use the correct formulas (4.2a), we would find

$$U = -\frac{1}{2} \int \frac{1}{4\pi E^2} \left[\left(\frac{\partial \Psi}{\partial x}\right)^2 + \left(\frac{\partial \Psi}{\partial y}\right)^2 + \left(\frac{\partial \Psi}{\partial z}\right)^2 \right] d\omega. \tag{4.23b}$$

[74] The formula (4.23a) can be transformed using integration by parts; since Maxwell denies the existence of surfaces of discontinuity,[19] it can be put in the form

$$U = \frac{1}{2} \int \Psi \left[\frac{\partial}{\partial x} \left(\frac{1}{4\pi E^2} \frac{\partial \Psi}{\partial x} \right) + \frac{\partial}{\partial y} \left(\frac{1}{4\pi E^2} \frac{\partial \Psi}{\partial y} \right) \right.$$
$$\left. + \frac{\partial}{\partial z} \left(\frac{1}{4\pi E^2} \frac{\partial \Psi}{\partial z} \right) \right] d\omega. \tag{4.24}$$

---

[18] J. Clerk Maxwell, SCIENTIFIC PAPERS, vol. I, p. 497.

[19] In this passage, Maxwell reasons always as if E^2 had the same value throughout all of space; but one can easily free his reasonings from this hypothesis.

Thence, by equalities (4.3) and equality (4.13b), Maxwell, affected by a sign error similar to that which affects Eqs. (4.2b), derives the equality

$$U = \frac{1}{2} \int \Psi e \, d\omega. \tag{4.25}$$

This would be achieved if, for the correct equality (4.23b), one applied the correct relation (4.12).

Maxwell also arrives at an expression of the electrostatic energy similar in form to the expression (3.7) that he admitted in his first theory. But, along the way, he met with equality (4.23a), which, once corrected of the sign error that affects the equations of the memoir: *On Physical Lines of Force*, takes the form (4.23b).

However, equality (4.23b) leads to a disturbing result.

The electrostatic energy of the system, zero in a depolarized system, would be negative in a polarized system; it would decrease because of the polarization. A set of dielectrics in the neutral state would be in an unstable state; once this state is disturbed, it would be polarized with ever-increasing intensity.

When Maxwell composed his memoir: *A Dynamical Theory of the Electromagnetic Field*, he resumed the equations given in the previous memoir, only after having cleared them of the sign errors that altered them. Therefore, the result that we just mentioned could appear. Is this the reason [75] why he, in this new work, changed the expression of the electrostatic energy? Still, instead of keeping, for the definition of this quantity, equality (4.22), he now defines this quantity by the equality[20]

$$U = \frac{1}{2} \int (Pf + Qg + Rh) d\omega. \tag{4.26}$$

In truth, this equality is not given here as a definition or a postulate, but arises from reasoning that we will reproduce:

Energy may be stored up in the field in a different way, namely, by the action of electromotive force in producing electric displacement. The work done by a variable electromotive force, P, in producing a variable displacement, f, is got by integrating

$$\int P \, df$$

from

$$P = 0$$

up to the given value of P.

Since $P = Kf$, ..., this quantity becomes

$$\int Kf \, df = \frac{1}{2} Kf^2 = \frac{1}{2} Pf.$$

---

[20] J. Clerk Maxwell, SCIENTIFIC PAPERS, vol. I, p. 563.

Hence the intrinsic energy of any part of the field, as existing in the form of electric displacement, is

$$\frac{1}{2}\int (Pf + Qg + Rh)d\omega.$$

It seems to us that this reasoning should rather justify the opposite conclusion and force Maxwell to retain the expression [76] of electric energy given by equality (4.22), which he adopted at the outset.

It seems quite clear that, in the reasoning above, P, Q, R must be regarded as components of an electromotive force *internal* to the system, and not as the components of an *exterior* electromotive force generated in the system by bodies that are foreign to it.

Indeed, we can notice, first, that Maxwell never decomposes the ensemble of bodies he studies into two groups, of which one is viewed as arbitrarily given, while the other, subject to the action of the first, experiences changes which the physicist analyzes. It seems rather that his calculations are applicable to the entire universe, likened to an isolated system, such that all the actions that he considers are internal actions.

Secondly, if, in the above reasoning, P, Q, R were the components of an external electromotive force, Maxwell should have added the components of the inner electromotive force arising from the very fact of the polarization of the dielectric medium; the omission of this last force would make his calculation incorrect.

We must therefore think that work evaluated by Maxwell is for him an *internal work*; but then this work is equivalent to a *decrease* and not to an increase in the internal energy, such that the conclusion of Maxwell should be reversed.

Maxwell, however, retains it and, in a field where the equilibrium is established, where there is, therefore,

$$P = -\frac{\partial \Psi}{\partial x}, \quad Q = -\frac{\partial \Psi}{\partial y}, \quad R = -\frac{\partial \Psi}{\partial z}, \tag{4.3}$$

he writes[21] Eq. (4.26) in the form

$$U = -\frac{1}{2}\int \left(\frac{\partial \Psi}{\partial x}f + \frac{\partial \Psi}{\partial y}g + \frac{\partial \Psi}{\partial z}h\right)d\omega$$

[77] which an integration by parts transforms into

$$U = \frac{1}{2}\int \Psi \left(\frac{\partial f}{\partial x} + \frac{\partial g}{\partial y} + \frac{\partial h}{\partial z}\right)d\omega$$

or, in virtue of equality (4.5),

$$U = -\frac{1}{2}\int \Psi e\, d\omega. \tag{4.27}$$

---

[21] J. Clerk Maxwell, SCIENTIFIC PAPERS, vol. I, p. 568.

4.6 On the Forces Exerted Between Two Small Charged Bodies

From the expression of the electrostatic energy, Maxwell will seek to deduce the laws of ponderomotive forces exerted on a charged system.

We first study this solution in the memoir: *On Physical Lines of Force*.[22]

The starting point is the expression of the electrostatic energy given by the formula (4.25).

Maxwell, who in the memoir in question will never consider surfaces of discontinuity, includes there no surface electrification; nevertheless, to avoid certain objections that may be made to the following considerations, it will be good to take account of such electrification and put the electrostatic energy in the form

$$U = \frac{1}{2} \int \Psi e \, d\omega + \frac{1}{2} \int \Psi E \, dS, \qquad (4.28)$$

the second integral extending over the charged surfaces.

We imagine that all of space is filled with a homogeneous dielectric; E^2 will be in all respects the same value.[23] [78]

The solid electric density will be given by the equality

$$\frac{1}{4\pi E^2} \Delta \Psi - e = 0, \qquad (4.16a)$$

which Maxwell should write, due to the sign error that affects equalities (4.2b),

$$\frac{1}{4\pi E^2} \Delta \Psi + e = 0. \qquad (4.16b)$$

On the other hand, on a point of a surface of discontinuity where the normal has two directions N_i, N_e, the surface density will be, according to the first equality (4.17), the value given by the equality

$$\frac{1}{4\pi E^2} \left(\frac{\partial \Psi}{\partial N_1} + \frac{\partial \Psi}{\partial N_e} \right) - E = 0, \qquad (4.29a)$$

which Maxwell should write

$$\frac{1}{4\pi E^2} \left(\frac{\partial \Psi}{\partial N_1} + \frac{\partial \Psi}{\partial N_e} \right) + E = 0. \qquad (4.29b)$$

A surface of discontinuity S_1 is supposed to separate the dielectric medium across a portion 1 that we regard as likely to be moved in this medium, like a solid in a liquid.

---

[22] J. Clerk Maxwell, SCIENTIFIC PAPERS, vol. I, p. 497, 498.

[23] The reader will easily avoid any confusion between the coefficient E^2 and the surface density E.

The function Ψ, which we will designate by Ψ_1, will be supposed to be harmonic throughout space, except in region 1; in this region there will be a solid density e_1. The surface S_1 may carry, in addition, the surface density E_1. The electrostatic energy of the system will be

$$U_1 = \frac{1}{2} \int \Psi_1 e_1 \, d\omega_1 + \frac{1}{2} \int \Psi_1 E_1 \, dS_1.$$

If body 1 is moved by causing its polarization, U_1 will remain invariable.

By a surface S_2, we likewise isolate another part 2 of the dielectric. Let Ψ_2 be a harmonic function outside of region 2; it corresponds to a solid density e_2 at any point of the [79] region 2 and to a solid density E_2 at all points of the surface S_2. If this electrification existed only in the medium, the electrostatic energy would be

$$U_2 = \frac{1}{2} \int \Psi_2 e_2 \, d\omega_2 + \frac{1}{2} \int \Psi_2 E_2 \, dS_2.$$

Imagine now that these two charged bodies exist simultaneously in the dielectric medium and that the function Ψ has the value $(\Psi_1 + \Psi_2)$. The electrification of each of the two bodies will be the same as if there were only one. As to the electrostatic energy of the system, it will obviously be according to equality (4.28):

$$U = \frac{1}{2} \int (\Psi_1 + \Psi_2) e_1 \, d\omega_1 + \frac{1}{2} \int (\Psi_1 + \Psi_2) E_1 \, dS_1$$
$$+ \frac{1}{2} \int (\Psi_1 + \Psi_2) e_2 \, d\omega_2 + \frac{1}{2} \int (\Psi_1 + \Psi_2) E_2 \, dS_2$$

or

$$U = U_1 + U_2 + \frac{1}{2} \int \Psi_2 e_1 \, d\omega_1 + \frac{1}{2} \int \Psi_2 e_1 \, dS_1$$
$$+ \frac{1}{2} \int \Psi_1 e_2 \, d\omega_2 + \frac{1}{2} \int \Psi_1 e_2 \, dS_2. \qquad (4.30)$$

But Green's theorem easily gives the equality

$$\int \Psi_1 \Delta \Psi_2 \, d\omega_2 + \int \Psi_1 \left(\frac{\partial \Psi_2}{\partial N_{2i}} + \frac{\partial \Psi_2}{\partial N_{2e}} \right) dS_2$$
$$= \int \Psi_2 \Delta \Psi_1 \, d\omega_1 + \int \Psi_2 \left(\frac{\partial \Psi_1}{\partial N_{1i}} + \frac{\partial \Psi_1}{\partial N_{1e}} \right) dS_2.$$

Whether we make use of Eqs. (4.16a) and (4.29a) or we make use of Eqs. (4.16b) and (4.29b), this equality can be written

$$\int \Psi_1 e_2 \, d\omega_2 + \int \Psi_1 E_2 \, dS_2 = \int \Psi_2 e_1 \, d\omega_1 + \int \Psi_2 E_1 \, dS_1$$

[80] and transforms equality (4.30) into

$$U = U_1 + U_2 + \int \Psi_2 e_1 \, d\omega_1 + \int \Psi_2 E_1 \, dS_1. \tag{4.31}$$

Leaving body 2 stationary, move body 1. U_1, U_2 remain invariable and U experiences an increase

$$\partial U = \partial \int \Psi_2 e_1 \, d\omega_1 + \partial \int \Psi_2 E_1 \, dS_1. \tag{4.32}$$

Maxwell notes that ∂U represents the work that should be carried out to move body 1 or, in other words, the *resistant work* generated by the actions of body 2 on body 1. The work *carried out* by these actions is therefore

$$-\partial U = -\partial \int \Psi_2 e_1 \, d\omega_1 - \partial \int \Psi_2 E_1 \, dS_1.$$

Suppose that body 1 is a very small body and that $\partial x_1, \partial y_1, \partial z_1$ are the components of the displacement of this body. Let

$$q_1 = \int e_1 \, d\omega_1 + \int E_1 \, dS_1 \tag{4.33}$$

be its total electric charge. We will have

$$-\partial U = -q_1 \left(\frac{\partial \Psi_2}{\partial x_1} \partial x_1 + \frac{\partial \Psi_2}{\partial y_1} \partial y_1 + \frac{\partial \Psi_2}{\partial z_1} \partial z_1 \right).$$

Body 2 therefore exerts on the small body 1 a force whose components are

$$\begin{cases} X_{21} = -q_1 \dfrac{\partial \Psi_2}{\partial x_1} = q_1 P_2, \\[2mm] Y_{21} = -q_1 \dfrac{\partial \Psi_2}{\partial y_1} = q_1 Q_2, \\[2mm] Z_{21} = -q_1 \dfrac{\partial \Psi_2}{\partial z_1} = q_1 R_2. \end{cases} \tag{4.34}$$

[81] In virtue of the equalities (4.16b) and (4.29b), we can write

$$\Psi_2 = \int \frac{E^2 e_2}{r} \, d\omega_2 + \int \frac{E^2 E_2}{r} \, dS_2$$

$$= E^2 \int \frac{e_2}{r} \, d\omega_2 + E^2 \int \frac{E_2}{r} \, dS_2.$$

If body 2 is very small, and referred to as

$$q_2 = \int e_2 \, d\omega_2 + \int E_2 \, dS_2,$$ (4.33b)

we will have its total electric charge at the point (x_1, y_1, z_1),

$$\Psi_2 = E^2 \frac{q_2}{r}.$$ (4.35)

Equalities (4.34) will then become

$$\begin{cases} X_{21} = E^2 \dfrac{q_1 q_2}{r^2} \dfrac{\partial r}{\partial x_1}, \\[2mm] Y_{21} = E^2 \dfrac{q_1 q_2}{r^2} \dfrac{\partial r}{\partial y_1}, \\[2mm] Z_{21} = E^2 \dfrac{q_1 q_2}{r^2} \dfrac{\partial r}{\partial z_1}. \end{cases}$$ (4.36)

They teach us that body 2 exerts on body 1 a *repulsive* force

$$F = E^2 \frac{q_1 q_2}{r^2}.$$ (4.37)

But this result is obtained by equalities (4.16b) and (4.29b), which are affected by a sign error.[24] If we make use of [82] equalities (4.16a) and (4.29a), where this sign error is corrected, we would find that equality (4.35) should be replaced by the equality

$$\Psi_2 = -E^2 \frac{q_2}{r},$$ (4.35b)

and body 2 would exert on body 1 an attractive force

$$A = E^2 \frac{q_1 q_2}{r^2}.$$ (4.37b)

This consequence, that would have certainly surprised Maxwell, will not be found in the memoir: *A Dynamical Theory of the Electromagnetic Field*, due to the change of sign suffered by the expression of the electrostatic energy.

In this memoir,[25] Maxwell deals very briefly with the mutual actions of charged bodies causes in directing the reader wishing to follow the details of the reasoning, to the theory of the magnetic forces that he had given.

---

[24] In fact, Maxwell did not write Eq. (4.16b), but Eq. (4.16a) [*op. cit.*, equality (123)]; but then he admits the expression (4.35) of Ψ_2 as if he had written Eq. (4.16b).

[25] J. Clerk Maxwell, SCIENTIFIC PAPERS, vol. I, pp. 566 à 568.

This reasoning is, moreover, led exactly according to the approach we have just described; only, instead of taking expression (66) of the electrostatic energy as a point of departure, he takes as a point of departure expression (4.27) of this energy or, better, expression

$$U = -\frac{1}{2} \int \Psi e \, d\omega - \frac{1}{2} \int \Psi E \, dS. \tag{4.38}$$

From this change of sign of the electrostatic energy, the replacement of equalities (4.34) by equalities

$$
\begin{cases}
X_{21} = q_1 \dfrac{\partial \Psi_2}{\partial x_1} = -q_1 P_2, \\[2mm]
Y_{21} = q_1 \dfrac{\partial \Psi_2}{\partial y_1} = -q_1 Q_2, \\[2mm]
Z_{21} = q_1 \dfrac{\partial \Psi_2}{\partial z_1} = -q_1 R_2.
\end{cases}
\tag{4.39}
$$

results. [83]

According to these equations, the *ponderomotive field* created by body 2 at the point (x, y, z) would have components $-P_2, -Q_2, -R_2$, while the *electromotive field* created by the same body, at the same point, would have the components P_2, Q_2, R_2; these two fields would therefore be equal, but *of contrary sense*. Maxwell, who wrote[26] equalities (4.39), does not stop at this paradoxical conclusion. Replacing[27] Ψ_2 by the expression

$$\Psi_2 = -\frac{K}{4\pi} \frac{q_2}{r}, \tag{4.40}$$

analogous to equality (4.35), he finds the equalities

$$
\begin{cases}
X_{21} = \dfrac{K}{4\pi} \dfrac{q_1 q_2}{r^2} \dfrac{\partial r}{\partial x_1}, \\[2mm]
Y_{21} = \dfrac{K}{4\pi} \dfrac{q_1 q_2}{r^2} \dfrac{\partial r}{\partial y_1}, \\[2mm]
Z_{21} = \dfrac{K}{4\pi} \dfrac{q_1 q_2}{r^2} \dfrac{\partial r}{\partial z_1},
\end{cases}
\tag{4.41}
$$

$$F = \frac{K}{4\pi} \frac{q_1 q_2}{r^2}, \tag{4.42}$$

analogous to equalities (4.36) and (4.37).

---

[26] *Loc. cit.*, p. 568, equalities (D).

[27] In reality, Maxwell wrote

$$\Psi_2 = \frac{K}{4\pi} \frac{q_2}{r},$$

[*loc. cit.*, equality (43)]; but this sign error is offset by a sign error in equality (44).

Maxwell thus achieves a law analogous to Coulomb's law, but on the condition of making the rather strange and singularly peculiar hypothesis that charged bodies have the same dielectric power as the medium between them.

Moreover, this conclusion is obtained, in the memoir: *On Physical Lines of Force*, only by means of a clerical sign error and, in the memoir: *A Dynamical Theory of the Electromagnetic Field*, it is deduced from an expression of the electrostatic energy whose sign is obviously wrong. [84]

4.7 On the Capacitance of a Capacitor

Another problem of electrostatics concerned Maxwell in the two memoirs that we analyzed in this chapter: it is the calculation of the capacitance of a capacitor.

We follow, first of all, the solution of this particular problem[28] in the memoir: *On Physical Lines of Force*.

Imagine a flat dielectric plate of thickness θ placed between two conductive plates 1 and 2. Maxwell admits that the function Ψ takes inside the conductive plate 1 the constant value, Ψ_1 and inside conductive plate 2 the invariable value Ψ_2; he implies that in the dielectric Ψ is a linear function of the distance to one of the armatures.

To calculate the electrical distribution on such a system, Maxwell made use, both for conductors and dielectrics, of Eq. (4.13b); he needs to join to it, to make his reasoning rigorous, the analogous equation for the surface charge of surfaces of discontinuity. He deduced that the charge is localized to the surfaces of separation of the armatures and the dielectric. For the surface of separation of armature 1 and the dielectric, the surface density will be

$$E = -\frac{1}{4\pi E^2} \frac{\partial \Psi}{\partial N_i},$$ (4.43)

N_i being the normal towards the interior of the dielectric.

Besides,

$$\frac{\partial \Psi}{\partial N_i} = \frac{\Psi_2 - \Psi_1}{\theta}.$$

So if S is the surface area of each frame in contact with the dielectric, armature 1 will carry a charge

$$Q = ES = \frac{S}{4\pi E^2} \frac{\Psi_1 - \Psi_2}{\theta}.$$ (4.44)

[85] Armature 2 will carry a charge equal and opposite in sign.

---

[28] J. Clerk Maxwell, SCIENTIFIC PAPERS, vol. I, p. 500.

Maxwell defines the capacitance of the capacitor by the formula

$$C = \frac{Q}{\Psi_1 - \Psi_2}. \tag{4.45}$$

Equality (4.44) will then give

$$C = \frac{1}{4\pi E^2} \frac{S}{\theta}, \tag{4.46}$$

which allows us to consider $\frac{1}{4\pi E^2}$ as the *specific inductive capacity* of the dielectric.

But this result was obtained by use of equality (4.43), tainted by the same sign error of equality (4.13b). If we were to make use of the correct equality

$$E = \frac{1}{4\pi E^2} \frac{\partial \Psi}{\partial N_i}, \tag{4.43b}$$

to which the first equality (4.7) would lead us, we would find for the *capacitance of the capacitor* the *negative* value

$$C = -\frac{1}{4\pi E^2} \frac{S}{\theta}. \tag{4.46b}$$

The sign error that affects equalities (4.2b) and, hence, so many equalities in the memoir: *On Physical Lines of Force*, disappeared in the memoir: *A Dynamical Theory of the Electromagnetic Field*. Does the theory of the capacitor that this memoir[29] contains therefore lead to the paradoxical result that a capacitor has a negative capacity? Rather than allowing himself to be dead-ended, Maxwell commits here a new sign error, [86] even in the memoir: *On Physical Lines of Force*, and he wrote[30]

$$\frac{\partial \Psi}{\partial x} = Kf,$$

while a few pages before, it was written[31]

$$P = Kf$$

and also[32]

$$P = -\frac{\partial \Psi}{\partial x}.$$

[87]

---

[29] J. Clerk Maxwell, SCIENTIFIC PAPERS, vol. I, p. 572.

[30] *Loc. cit.*, p. 572, equality (48).

[31] *Loc. cit.*, p. 560, equalities (E).

[32] *Loc. cit.*, p. 568.

Chapter 5
The Third Electrostatics of Maxwell

5.1 Essential Difference Between the Second and the Third Electrostatics of Maxwell

The sign errors that we just pointed out can only hide the inevitable contradiction which the theory of a given capacitor faces in the second electrostatics of Maxwell.

In this electrostatics, the electric density e is taken into account. This density arises because the electrification of some polarized corpuscle, of which Maxwell admits the existence, like Faraday and Mossotti, is not exactly neutralized by the electrification of the neighboring corpuscles; this density is the analogue of the fictional density that Poisson taught us to substitute for the magnetization of a piece of iron. In any case, there is no question of an electric density other than that one. Maxwell takes into account an electrification that is not reducible to the polarization of dielectrics, of an electrification proper to conductive bodies. What is clearer, for example, than the following passage[1] that we read in the memoir: *A Dynamical Theory of the Electromagnetic Field*?

Electric Quantity

Let e represent the quantity of free positive electricity contained in unit of volume at any part of the field, then, since this arises from the electrification of the different parts of the field not neutralizing [88] each other, we may write the *equation of free electricity*:

$$e + \frac{\partial f}{\partial x} + \frac{\partial g}{\partial y} + \frac{\partial h}{\partial z} = 0.$$

Admitting this fundamental principle of the theories of Maxwell, we resume the study of a flat capacitor made of two conductive sheets 1 and 2 that a dielectric separates.

---

[1] J. Clerk Maxwell, Scientific Papers, vol. I, p. 561.

© Springer International Publishing Switzerland 2015
P.M.M. Duhem, *The Electric Theories of J. Clerk Maxwell*,
Boston Studies in the Philosophy and History of Science 314,
DOI 10.1007/978-3-319-18515-6_5

Suppose that the inner side of plane 1 is positively charged and the inner side of plane 2 is negatively charged; within the dielectric plane, the electromotive field is directed from plane 1 to plane 2.

If, according to the sign error committed by Maxwell in his memoir: *On Physical Lines of Force* and contained in the part of the memoir: *A Dynamical Theory of the Electromagnetic Field* where he considers the theory of the capacitor, we suspected the displacement headed in the opposite direction of the electromotive field, the displacement would be within the dielectric plane, directed from conductor 2 toward conductor 1.

But, except where we just pointed out, Maxwell has never reproduced this opinion in his writings after the memoir: *On Physical Lines of Force*. Everywhere, he admits that the displacement, proportional to the electromotive force, is directed parallel to it.

He wrote in 1868[2]:

If we admit that the energy of the system so electrified resides in the polarized dielectric, we must also admit that within the dielectric there is a displacement of electricity in the direction of the electromotive force...

He repeated in his *Treatise*[3]:

The displacement is in the same direction as the force, and is numerically equal to the intensity [89] multiplied by $\frac{K}{4\pi}$, where K is the specific inductive capacity of the dielectric.

He said further on[4]:

In this treatise, static electric induction is measured by what we have called the *electric displacement*, a directed quantity or vector which we have denoted by $\mathfrak{D}$, and its components by f, g, h.

In isotropic substances, the displacement is in the same direction as the electromotive force which produces it, and is proportional to it, at least for small values of this force. This may be expressed by the equation:

$$\text{Equation of Electric Displacement, } \mathfrak{D} = \frac{K}{4\pi} \mathfrak{E},$$

where K is the dielectric capacity of the substance.

---

[2] J. Clerk Maxwell, *On a Method of Making a Direct Comparison of Electrostatic with Electromagnetic Force: With a Note on the Electromagnetic Theory of Light* (Read at the Royal Society of London on 18 June 1868. PHILOSOPHICAL TRANSACTIONS, vol. Society of London on 18 June 1868. PHILOSOPHICAL TRANSACTIONS, vol. CLVIII.—SCIENTIFIC PAPERS, vol. II, p. 139).

[3] J. Clerk Maxwell, *A Treatise on Electricity and Magnetism*; Oxford, 1873, vol. I, p. 63.—*Traité d'Électricité et de Magnétisme*, translated from English in the 2nd edition, by G. Seligmann-Lui; Paris, 1885–1887; Volume I, p. 73 [65].—We will cite the Treatise of Maxwell according to the French translation whenever no changes have been made to the 1st English edition.

[4] *Treatise...*, vol. II, p. 287 [232].

If we designate, with Maxwell, by P, Q, R the components of the electromotive force $\mathfrak{E}$, the preceding symbolic equality will be equal to the three equalities[5]

$$f = \frac{K}{4\pi}P, \quad g = \frac{K}{4\pi}Q, \quad h = \frac{K}{4\pi}R. \tag{5.1}$$

Finally, in the book which Maxwell was preparing the publication shortly before his death, in one of the chapters that are entirely of his hand, we read[6]:

According to the theory adopted in this book, when an electromotive force acts on a dielectric it causes the electricity to be displaced within it in the direction of the electromotive force, the amount of the displacement being proportional to the electromotive force, but depending also on the nature of the dielectric...

Therefore, if a dielectric plate is between the two [90] armatures of a condenser of which one is electrified positively and one negatively, the displacement will be, at each point, directed from the positive armature to the negative armature. Maxwell admits this law without hesitation; he writes in his *Note on the Electromagnetic Theory of Light*[7]:

When a dielectric is acted on by electromotive force it experiences what we may call electric polarization. If the direction of the electromotive force is called positive, and if we suppose the dielectric bounded by two conductors, A on the negative, and B on the positive side, then the surface of the conductor A is positively electrified, and that of B negatively.

If we admit that the energy of the system so electrified resides in the polarized dielectric, we must also admit that within the dielectric there is a displacement of electricity in the direction of the electromotive force...

He repeats in his great *Treatise*:

[8]The positive electrification of A and the negative electrification of B will produce a certain electromotive force acting from A towards B in the dielectric stratum, and this will produce an electric displacement from A towards B within the dielectric.

He later wrote[9]:

The displacements across any two sections of the same tube of displacement are equal. At the beginning of each unit tube of displacement there is a unit of positive electricity, and at the end of the tube there is a unit of negative electricity.

---

[5]The comparison of the equalities (5.1) with equalities (4.4) shows that the quantity $\frac{K}{4\pi}$ introduced here by Maxwell is what is designated by $\frac{1}{K}$ in his memoir *A Dynamical Theory of the Electromagnetic Field*.

[6]J. Clerk Maxwell, *An Elementary Treatise on Electricity*, edited by W. Garnett.—*Traité élémentaire d'Électricité*, translated from English by Gustave Richard. Paris 1884, p. 141 [108].

[7]J. Clerk Maxwell, SCIENTIFIC PAPERS, vol. II, p. 339.

[8]J. Clerk Maxwell, *Treatise on Electricity and Magnetism*, t. I, p. 71 [63].

[9]J. Clerk Maxwell, *An Elementary Treatise on Electricity*, p. 71 [53].

What exact meaning does Maxwell attribute, in his final works, to the words *electric displacement*?

In the memoir *A Dynamical Theory of the Electromagnetic Field*, where he introduced this term for the first time, Maxwell, we saw, was inspired by Mossotti. For Mossotti, the electromotive force, meeting one of the corpuscles of which the dielectric body is comprised, drives out the ethereal fluid from the parts of the surface where it enters the corpuscle, to accumulate on the regions where it was released. The thought of Maxwell, in the two memoirs we analyzed in the previous chapter, is fully consistent with that of Mossotti. Is it the same in his most recent writings? [91]

We cannot doubt; the displacement remains, for Maxwell, a driving of the positive electricity that the electromotive force produced in its own direction, a driving that is limited to each small portion of the dielectric:

> The electric polarization of an elementary portion of a dielectric[10] is a forced state into which the medium is thrown by the action of electromotive force, and which disappears when that force is removed. We may conceive it to consist in what we may call an electrical displacement, produced by the electromotive intensity. When the electromotive force acts on a conducting medium it produces a current through it, but if the medium is a non-conductor or dielectric, the current cannot flow through the medium, but the electricity is displaced within the medium in the direction of the electromotive intensity, the extent of this displacement depending on the magnitude of the electromotive intensity, so that if the electromotive intensity increases or diminishes, the electric displacement increases and diminishes in the same ratio.
>
> The amount of the displacement is measured by the quantity of electricity which crosses unit of area, while the displacement increases from zero to its actual amount. This, therefore, is the measure of the electric polarization.

The following passage is, if possible, even more formal[11]:

> To make our conception of what takes place more precise, let us consider a single cell belonging to a tube of induction proceeding from a positively electrified body, the cell being bounded by two consecutive equipotential surfaces surrounding the body.
>
> We know that there is an electromotive force acting outwards from the electrified body. This force, if it acted on a conducting medium, would produce a current of electricity which [92] would last as long as the force continued to act. The medium however is a non-conducting or dielectric medium, and the effect of the electromotive force is to produce what we may call *electric displacement*, i.e., the electricity is forced outwards in the direction of the electromotive force, but its condition when so displaced is such that, as soon as the electromotive force is removed, the electricity resumes the position which it had before displacement.

The idea that Maxwell indicates, in his final works, by these words: *electric displacement*, is therefore consistent with what he means when speaking about the same words in his first memoirs, starting with the conception of Mossotti, with the

---

[10] J. Clerk Maxwell, *Treatise on Electricity and Magnetism*, t. I, p. 69 [61–62].

[11] J. Clerk Maxwell, *An Elementary Treatise on Electricity*, p. 61 [48–49].

theory of magnetic induction that the genius of Poisson created. Moreover, Maxwell carefully noted this agreement[12]:

> Since, as we have seen, the theory of direct action at a distance is mathematically identical with that of action by means of a medium, the actual phenomena may be explained by the one theory as well as by the other...[13] Thus, Mossotti has deduced the mathematical theory of dielectrics from the ordinary theory of attraction merely by giving an electric instead of a magnetic interpretation to the symbols in the investigation by which Poisson has deduced the theory of magnetic induction from the theory of magnetic fluids. He assumes the existence within the dielectric of small conducting elements, capable of having their opposite surfaces oppositely electrified by induction, but not capable of losing or gaining electricity on the whole, owing to their being insulated from each other by a non-conducting medium. This theory of dielectrics is consistent with the laws of electricity, and may be actually true. If it is true, the specific inductive capacity of a dielectric may be greater, but cannot be less, than that of a vacuum. No instance has yet been found of a dielectric having an inductive capacity less than that of a vacuum, but if such should be discovered, Mossotti's physical theory must be abandoned, although his formulas [93] would all remain exact, and would only require us to alter the sign of a coefficient.
>
> [In the theory that I propose to develop, the mathematical methods are founded on the smallest possible number of hypotheses][14]; in many parts of physical science, equations of the same form are found applicable to phenomena which are certainly of quite different natures, as, for instance, electric induction through dielectrics, conduction through conductors, and magnetic induction. In all these cases the relation between the force and the effect produced is expressed by a set of equations of the same kind, so that when a problem in one of these subjects is solved, the problem and its solution may be translated into the language of the other subjects and the results in their new form will still be true.

From all these citations, a consequence seems to flow logically, between the components f, g, h of displacement and the solid or surface electrical densities e, E, we must establish the relations

$$e + \frac{\partial f}{\partial x} + \frac{\partial g}{\partial y} + \frac{\partial h}{\partial z} = 0, \tag{4.5}$$

$$\begin{aligned} E + f_1 \cos(N_1, x) + g_1 \cos(N_1, y) + h_1 \cos(N_1, z) \\ + f_2 \cos(N_2, x) + g_2 \cos(N_2, y) + h_2 \cos(N_2, z) = 0. \end{aligned} \tag{4.6}$$

These equations, indeed, agree with what Maxwell said on electric displacement; they are among the essential formulas of Mossotti's theory, which Maxwell declared mathematically identical to his own. They are, moreover, in this theory, the transposition of equations that Poisson introduced into the theory of magnetic induction and which Maxwell[15] keeps in his exposition of this latter theory; finally, Maxwell adopted them in his early writings.

---

[12] J. Clerk Maxwell, *Treatise on Electricity and Magnetism*, t. 1, p. 74 [66–67].

[13] [Duhem omits the rest of the sentence: "provided suitable hypotheses be introduced when any difficulty occurs"].

[14] [The French translation adds this, which is not found here in the original English.].

[15] J. Clerk Maxwell, *Treatise on Electricity and Magnetism*, t. II, p. 11 [10].

We are again led to recognize that the ideas of Maxwell lead logically to equalities (4.5) and (4.6) by analysing what he says about *displacement currents*. [94]

> The variations of electric displacement evidently constitute electric currents.[16] These currents, however, can only exist during the variation of the displacement...
>
> One of the chief peculiarities of this treatise[17] is the doctrine which it asserts, that the true electric current $\mathfrak{C}(u, v, w)$ that on which the electromagnetic phenomena depend, is not the same thing as $\mathfrak{K}(p, q, r)$, the current of conduction, but that the time-variation of $\mathfrak{D}$, the electric displacement, must be taken into account in estimating the total movement of electricity, so that we must write,

$$\text{Equation of true currents, } \mathfrak{C} = \mathfrak{K} + \frac{\partial \mathfrak{D}}{\partial t}$$

or, in terms of the components,

$$u = p + \frac{\partial f}{\partial t},$$

$$v = q + \frac{\partial g}{\partial t},$$

$$w = r + \frac{\partial h}{\partial t}.$$

Thus, at any point of a non-conducting dielectric whose polarization varies, a displacement current varies, whose components are

$$p' = \frac{\partial f}{\partial t}, \quad q' = \frac{\partial g}{\partial t}, \quad r' = \frac{\partial h}{\partial t}. \tag{5.2}$$

Yet,

> whatever electricity may be,[18] and whatever we may understand by the movement of electricity, the phenomenon which we have called *electric displacement* is a movement of electricity in the same sense as the transference of a definite quantity of electricity...

Either the sentence does not mean anything, or it requires that the components [95] p', q', r' of the displacement current are linked to electric densities e, E by the equations of continuity

$$\frac{\partial p'}{\partial x} + \frac{\partial q'}{\partial y} + \frac{\partial r'}{\partial z} + \frac{\partial e}{\partial t} = 0,$$

$$\frac{\partial E}{\partial t} + p_1' \cos(N_1, x) + q_1' \cos(N_1, y) + r_1' \cos(N_1, z)$$
$$+ p_2' \cos(N_2, x) + q_2' \cos(N_2, y) + r_2' \cos(N_2, z) = 0,$$

---

[16] J. Clerk Maxwell, *Treatise on Electricity and Magnetism*, t. I, p. 69 [62].

[17] *Ibid.*, t. II, p. 288 [232–233].

[18] *Ibid.*, t. I, p. 73 [66].

which can be written, in virtue of equalities (5.2),

$$\frac{\partial}{\partial t}\left(\frac{\partial f}{\partial x} + \frac{\partial g}{\partial y} + \frac{\partial h}{\partial z} + e\right) = 0,$$

$$\frac{\partial}{\partial t}[E + f_1 \cos(N_1, x) + g_1 \cos(N_1, y) + h_1 \cos(N_1, z)$$
$$+ f_2 \cos(N_2, x) + g_2 \cos(N_2, y) + h_2 \cos(N_2, z)] = 0.$$

Integrated between the times when the system was in the neutral state and the current state, these equations give Eqs. (4.5) and (4.6); this reasoning is, moreover, given by Maxwell in his memoir: *On Physical Lines of Force.*

Examine the consequences of these equations and, in particular, Eq. (4.6); apply it to the surface of separation of a dielectric 1 and a non-polarizable conductor 2. Letting the displacement (f_2, g_2, h_2) be zero in this latter medium, Eq. (4.6) reduces to

$$E + f_1 \cos(N_1, x) + g_1 \cos(N_1, y) + h_1 \cos(N_1, z) = 0. \qquad (5.3)$$

The terminal surface of the dielectric is electrified negatively at the points where the direction of displacement or, what amounts to the same, the direction of the electromotive force, enters the dielectric; it is positively electrified at the points where this same direction exits the dielectric.

Applying this proposition, which naturally follows from the [96] principles laid down by Maxwell, to our dielectric plate between two charged conductors, the one of positive electricity, the other of negative electricity, we obtain the following conclusion:

The side of the dielectric which is in contact with the positively electrified conductor carries negative electricity; the side that is in contact with the negatively electrified conductor carries positive electricity. It is therefore impossible to identify the electrical charge that a conductor carries with the charge taken by the adjacent dielectric.

Will Maxwell therefore abandon the hypothesis, implied in his early writings, that there is no such thing as the proper electrification of the conductive bodies; that, alone, the polarization of the dielectric media is a real phenomenon, producing, by the apparent electrification to which it is equivalent, the effects that the old theories attribute to electrical charges spread over conductive bodies? Quite to the contrary; he outlines more clearly this hypothesis and affirms its legitimacy:

He said[19]:

We may conceive the physical relation between the electrified bodies, either as the result of the state of the intervening medium, or as the result of a direct action between the electrified bodies at a distance.

---

[19] J. Clerk Maxwell, *Treatise on Electricity and Magnetism*, t. I, p. 67 [59–60].

...If we calculate on this hypothesis the total energy residing in the medium, we shall find it equal to the energy due to the electrification of the conductors on the hypothesis of direct action at a distance. Hence the two hypotheses are mathematically equivalent.

In the interior of the medium[20] where the positive end of one cell is in contact with the negative end of the next, these two electrifications exactly neutralise each other, but where the dielectric medium is bounded by a conductor, the electrification is no longer neutralised, but constitutes the observed electrification at the surface of the conductor.

According to this view of electrification, we must regard electrification as a property of the dielectric medium rather than of the conductor which is bounded by it. [97]

In the case of the Leyden jar[21] of which the inner coating is charged positively, any portion of the glass will have its inner side charged positively and its outer side negatively. If this portion be entirely in the interior of the glass, its surface charge will be neutralized by the opposite charge of the parts in contact with it, but if it be in contact with a conducting body, which is incapable of maintaining in itself the inductive state, the surface charge will not be neutralized, but will constitute that apparent charge which is commonly called the *Charge of the Conductor*.

The charge therefore at the bounding surface of a conductor and the surrounding dielectric, which on the old theory was called the charge of the conductor, must be called in the theory of induction the *surface charge of the surrounding dielectric*.

According to this theory, all charge is the residual effect of the polarization of the dielectric.

Since Maxwell formally maintains this hypothesis, how will he remove the contradiction that we have reported? The simplest way: in Eqs. (4.5) and (4.6), which render this contradiction glaring, he will change the sign of e and E and write[22]

$$e = \frac{\partial f}{\partial x} + \frac{\partial g}{\partial y} + \frac{\partial h}{\partial z},$$ (5.4)

$$E = f_1 \cos(N_1, x) + g_1 \cos(N_1, y) + h_1 \cos(N_1, z)$$
$$+ f_2 \cos(N_2, x) + g_2 \cos(N_2, y) + h_2 \cos(N_2, z).$$ (5.5)

Equality (5.3), which made known the surface charge of a dielectric in contact with a conductor—i.e. in the hypothesis of Maxwell, the charge of the the the conductor itself—will be replaced by equality

$$E = f_1 \cos(N_1, x) + g_1 \cos(N_1, y) + h_1 \cos(N_1, z).$$ (5.6)

[20] J. Clerk Maxwell, *An Elementary Treatise on Electricity*, p. 63 [49].
[21] J. Clerk Maxwell, *Treatise on Electricity and Magnetism*, [t. I,] p. 175 [155].
[22] *Ibid.*, [t. II,] p. 289 [233].

[98] The charge will be positive where the direction of displacement or the electromotive field penetrates inside the dielectric and negative where the direction of displacement or the electromotive field exits the dielectric.

In the case of the charged conductor[23] let us suppose the charge to be positive, then if the surrounding dielectric extends on all sides beyond the closed surface there will be electric polarization, accompanied with displacement from within outwards all over the closed surface, and the surface-integral of the displacement taken over the surface will be equal to the charge on the conductor within.

How should the elementary masses of a dielectric be polarized, if it is desired that the electrification in opposite sense of their two ends agrees with equalities (5.4), (5.5) and (5.6)?

Let us take the example of a planar dielectric plate placed between two conductive plates. It is assumed that within the plate the electrical charges that are at the two ends of a molecule are exactly neutralized by the charges of the molecule that precedes it and by the molecule that follows it. Only the electric charge of the molecules at the ends produces appreciable effects.

The face of the dielectric through which the electromotive force enters into the medium manifests a state of electrification; it is due to the charge that the molecules of the first layer take in one of their extremities through which the electromotive force penetrates them. The face of the dielectric through which the electromotive force exits the medium also manifests a state of electrification. It is caused by the charge that, in one of their extremities through which the electromotive force leaves, the molecules of the last layer take. However, according to the propositions that Maxwell comes to state, the first electrification is positive, the last negative. Therefore, *when an electromotive force meets a dielectric molecule, it polarizes it; the end of the molecule through which the electromotive force enters is responsible for* POSITIVE *electricity; the end of the molecule through which the electromotive force is released is responsible for* NEGATIVE *electricity.*

This is the proposition[24]—contrary to what Coulomb and Poisson admitted in their study of magnetism, contrary to Faraday and Mossotti in their study of dielectrics, and contrary to the view he himself professed in his early writings—that Maxwell formally outlines in his final treaties.

That the surface of any elementary portion into which we may conceive[25] the volume of the dielectric divided must be conceived to be charged so that the surface-density at any point of the surface is equal in magnitude to the displacement through that point of the surface *reckoned inwards*. If the displacement is in the positive direction, the surface of the element will be charged negatively on the positive side of the element, and positively on the negative side. These superficial charges will in general destroy one another when consecutive

---

[23] J. Clerk Maxwell, *Treatise on Electricity and Magnetism*, t. I, p. 72 [64].

[24] I do not believe that any physicist has paid attention to the paradoxical nature of this proposition of Maxwell before H. Hertz exposed it in a particularly clear and striking form (H. Hertz, GESAMMELTE WERKE, Bd. II: *Untersuchungen über die Aushreitung der elektrischen Kraft: Einleitende Uebersicht* [English translation [99]: Hertz (1893)], p. 27).

[25] J. Clerk Maxwell, *Treatise on Electricity and Magnetism*, t. I, p. 73 [65–66].

elements are considered, except where the dielectric has an internal charge, or at the surface of the dielectric.

In the case of the Leyden jar[26] of which the inner coating is charged positively, any portion of the glass will have its inner side charged positively and its outer side negatively.

The displacement[27] across any section of a unit tube of induction is one unit of electricity and the direction of the displacement is that of the electromotive force, namely, from places of higher to places of lower potential.

Besides the electric displacement within the cell we have to consider the state of the two ends of the cell which are formed by the equipotential surfaces. We must suppose that in every cell the end formed by the surface of higher potential is coated with one unit of positive electricity, the opposite [100] end, that formed by the surface of lower potential, being coated with one unit of negative electricity.

In the *Treatise on Electricity and Magnetism* as in the *Elementary Treatise on Electricity*, a few pages, sometimes a few lines only, separate the passages that we just quoted from statements such as these:

...the effect of the electromotive force[28] is to produce what we may call *electric displacement*, i.e. the electricity is forced outwards in the direction of the electromotive force...

When induction[29] is transmitted through a dielectric, there is in the first place a displacement of electricity in the direction of the induction. For instance, in a Leyden jar, of which the inner coating is charged positively and the outer coating negatively, the displacement of positive electricity in the substance of the glass is from within outwards.

When the electromotive force acts on a conducting medium[30] it produces a current through it, but if the medium is a non-conductor or dielectric, the current cannot flow through the medium, but the electricity is displaced within the medium in the direction of the electromotive intensity...

That whatever electricity may be,[31] and whatever we may understand by the movement of electricity, the phenomenon which we have called *electric displacement* is a movement of electricity in the same sense as the transference of a definite quantity of electricity through a wire is a movement of electricity...

Either this language does not mean anything, or it means the following: when an electromotive force is an elementary part of the dielectric, the state of electrical neutrality of this part is disturbed; *electricity moves in the direction of the electromotive force; it accumulates in excess at the end where the electromotive force exits out of the particle, so that this end is electrified* POSITIVELY, *while it leaves the end where the electromotive force enters into the particle, and this end is electrified* NEGATIVELY. [101]

---

[26] *Ibid.*, t. I, p. 175 [155].

[27] J. Clerk Maxwell, *An Elementary Treatise on Electricity*, p. 63 [49].

[28] J. Clerk Maxwell, *An Elementary Treatise on Electricity*, p. 62 [49].

[29] J. Clerk Maxwell, *Treatise on Electricity and Magnetism*, t. I, p. 174 [154].

[30] *Ibid.*, p. 69 [61–62].

[31] *Ibid.*, p. 73 [66].

How can two so clearly contradictory propositions occur at the same time for the mind of Maxwell and, both at once, cause his assent? This is a strange problem of scientific psychology that we deliver to the meditations of the reader.

5.2 Development of the Third Electrostatics of Maxwell

If one condemns this first contradiction, if one accepts equalities (5.4), (5.5) and (5.6), the equations of the third electrostatics of Maxwell unfold, through the course of his *Treatise*, free from the continual changes of sign that interrupted the course of the second electrostatics.

If P, Q, R are the components of the electromotive force, the components f, g, h of displacement are given by equalities[32]

$$f = \frac{K}{4\pi}P, \quad g = \frac{K}{4\pi}Q, \quad h = \frac{K}{4\pi}R, \tag{5.7}$$

where K is the specific inductive capacity of the dielectric.

The *electrostatic energy* of the medium is given by the following proposition[33]:

The most general expression for the electric energy of the medium per unit of volume is half the product of the electromotive intensity and the electric polarization multiplied by the cosine of the angle between their directions.

In all fluid dielectrics the electromotive intensity and the electric polarization are in the same direction…

For these latter bodies,[34] the electrostatic energy is therefore

$$U = \frac{1}{2}\int (Pf + Qg + Rh)d\omega. \tag{5.8}$$

[102] It is also for the same bodies that Eq. (5.7) are valid, thereby giving the electrostatic energy these two other expressions[35]:

$$U = \frac{1}{8\pi}\int K(P^2 + Q^2 + R^2)d\omega, \tag{5.9}$$

$$U = 2\pi \int \frac{f^2 + g^2 + h^2}{K} d\omega. \tag{5.10}$$

---

[32] J. Clerk Maxwell, *Treatise on Electricity and Magnetism*, t. I, p. 73 [65]; t. II, p. 287 [232].
[33] *Ibid.*, t. I, p. 67 [60].
[34] *Ibid.*, t. I, p. 176 [156]; t. II, p. 304 [246].
[35] J. Clerk Maxwell, *Treatise on Electricity and Magnetism*, t. I, p. 176 [156].

where the system is in electrical equilibrium, the laws of electrodynamics show that there is a certain function $\Psi(x, y, z)$ such that[36]

$$P = -\frac{\partial \Psi}{\partial x}, \quad Q = -\frac{\partial \Psi}{\partial y}, \quad R = -\frac{\partial \Psi}{\partial z}. \tag{5.11}$$

The expressions (5.8) and (5.9) of the internal energy of a system in equilibrium can then be written[37]:

$$U = -\frac{1}{2} \int \left(\frac{\partial \Psi}{\partial x} f + \frac{\partial \Psi}{\partial y} g + \frac{\partial \Psi}{\partial z} h \right) d\omega, \tag{5.12}$$

$$U = \frac{1}{8\pi} K \int \left[\left(\frac{\partial \Psi}{\partial x} \right)^2 + \left(\frac{\partial \Psi}{\partial y} \right)^2 + \left(\frac{\partial \Psi}{\partial z} \right)^2 \right] d\omega. \tag{5.13}$$

An integration by parts can turn equality (5.12) into equality

$$\begin{aligned}
U = \frac{1}{2} \int \Psi &\left(\frac{\partial f}{\partial x} + \frac{\partial g}{\partial y} + \frac{\partial h}{\partial z} \right) d\omega \\
+ \frac{1}{2} \int \Psi &[f_1 \cos(N_1, x) + g_1 \cos(N_1, y) + h_1 \cos(N_1, z) \\
&+ f_2 \cos(N_2, x) + g_2 \cos(N_2, y) + h_2 \cos(N_2, z)] dS,
\end{aligned}$$

[103] the last integral extending to various surfaces of discontinuity.

By the use of formulas (5.4), (5.5) and (5.6), this equality becomes[38]

$$U = \frac{1}{2} \int \Psi e \, d\omega + \frac{1}{2} \int \Psi E \, dS. \tag{5.14}$$

Moreover, in virtue of equalities (5.7) and (5.11), we have

$$f = -\frac{K}{4\pi} \frac{\partial \Psi}{\partial x}, \quad g = -\frac{K}{4\pi} \frac{\partial \Psi}{\partial y}, \quad h = -\frac{K}{4\pi} \frac{\partial \Psi}{\partial z}, \tag{5.15}$$

and relations (5.4) and (5.5) become

$$\frac{\partial}{\partial x} \left(K \frac{\partial \Psi}{\partial x} \right) + \frac{\partial}{\partial y} \left(K \frac{\partial \Psi}{\partial y} \right) + \frac{\partial}{\partial z} \left(K \frac{\partial \Psi}{\partial z} \right) + 4\pi e = 0, \tag{5.16}$$

$$K_2 \frac{\partial \Psi}{\partial N_1} + K_2 \frac{\partial \Psi}{\partial N_2} + 4\pi E = 0. \tag{5.17}$$

---

[36] *Ibid.*, t. II, p. 274 [221], equation (B).
[37] *Ibid.*, t. II, p. 303 [246].
[38] J. Clerk Maxwell, *Treatise on Electricity and Magnetism* t. I, p. 108 [96]; t. II, p. 303 [246].

Maxwell introduced these equalities[39] in his *Treatise* not by the above reasoning, but by a strange and little-known analogy between these equalities and the relations of Poisson

$$\frac{\partial^2 V}{\partial x^2} + \frac{\partial^2 V}{\partial y^2} + \frac{\partial^2 V}{\partial z^2} + 4\pi e = 0, \tag{5.18a}$$

$$\frac{\partial V}{\partial N_1} + \frac{\partial V}{\partial N_2} + 4\pi E = 0, \tag{5.19a}$$

which satisfy the function

$$V = \int \frac{e}{r} d\omega + \int \frac{E}{r} d\omega. \tag{5.20a}$$

[104] In a note[40] added to the French translation of the *Treatise* of Maxwell, Potier has already done justice to this approximation; it is good to emphasize what of it is fallacious.

Equalities (5.18a) and (5.19a) are purely algebraic consequences of the analytical form of the function V, as given by equality (5.20a); on the contrary, the analytical form of the function Ψ is unknown, and equalities (5.16) and (5.17) are the result of physical hypotheses.

5.3 A Return to the First Electrostatics of Maxwell

The equations we have just written offer a profound analogy with the equations which guides the heat conductivity theory. In his *Treatise on Electricity and Magnetism*, Maxwell does not resume this analogy, which had been the starting point of his research on dielectric media; but he insists on it in his *Elementary Treatise on Electricity*.[41] And indeed, one easily passes from the formulas of the theory of heat, given in Chap. 3, to the formulas that Maxwell gives in his *Treatise on Electricity and Magnetism* if, between the quantities that appear in these formulas, the following correspondence table is established:

Theory of heat	Electrostatic
T, temperature	Ψ
u, v, w, components of heat flux	f, g, h, components of the electric displacement
k, coefficient of heat conductivity	$\frac{K}{4\pi}$, specific inductive capacity K
j, intensity of a solid heat source	e, solid electrical density
J, surface intensity of a heat source	E, electric surface density

---

[39] *Ibid.*, t. I, p. 104 [94].

[40] J. Clerk Maxwell, *Treatise on Electricity and Magnetism*, t. I, p. 106.

[41] J. Clerk Maxwell, *Treatise on Electricity and Magnetism*, p. [51,] [§] 64.

[105] Therefore, equalities (3.1), (3.2) and (3.3) turn into equalities (5.15), (5.16) and (5.17).

But in developing his first electrostatics, Maxwell, we have seen, admitted that the function Ψ is expressed analytically, as the potential function V used in classical electrostatics, by the formula

$$\Psi = \int \frac{e}{r} d\omega + \int \frac{E}{r} dS.$$

In his *Treatise on Electricity and Magnetism*,[42] on the contrary, he puts his reader on guard against this confusion; by *apparent distribution of electricity* means a distribution whose solid density e' and the surface density E' would make the function Ψ known by the formula

$$\Psi = \int \frac{e'}{r} d\omega + \int \frac{E'}{r} dS. \tag{5.20b}$$

According to the theorems of Poisson, we would then have equalities

$$\frac{\partial^2 V}{\partial x^2} + \frac{\partial^2 V}{\partial y^2} + \frac{\partial^2 V}{\partial z^2} + 4\pi e = 0, \tag{5.18b}$$

$$\frac{\partial \Psi}{\partial N_1} + \frac{\partial \Psi}{\partial N_2} + 4\pi E' = 0. \tag{5.19b}$$

By comparing these equalities to equalities (5.16) and (5.17), we see that the *densities* e', E' cannot be equal to densities e, E. In particular, equalities (5.16) and (5.18b) give,

$$4\pi(Ke' - e) = \frac{\partial K}{\partial x}\frac{\partial \Psi}{\partial x} + \frac{\partial K}{\partial y}\frac{\partial \Psi}{\partial y} + \frac{\partial K}{\partial z}\frac{\partial \Psi}{\partial z}. \tag{5.21}$$

[106] Equalities (5.17) and (5.19b) give[43]

$$\begin{cases} 4\pi(K_2E' - E) = (K_1 - K_2)\dfrac{\partial \Psi}{\partial N_1}, \\[2mm] 4\pi(K_1E' - E) = (K_2 - K_1)\dfrac{\partial \Psi}{\partial N_2}. \end{cases} \tag{5.22}$$

The place would be here, it seems, to judge this electrostatics of Maxwell and see if it can agree with known laws; but we lack one thing to complete this discussion; this thing is the concept of *displacement current*, which belongs to electrodynamics.

---

[42] J. Clerk Maxwell, *Treatise on Electricity and Magnetism*, t. I, p. 104 [94–95].

[43] These equalities (5.22) are, in all editions of the *Treatise* of Maxwell, replaced by erroneous equalities. In the French translation, the term Ke' of equality (5.21) is replaced by e'; this error is not in the first English edition.

Part II
The Electrodynamics of Maxwell

Chapter 6
Conduction Current and Displacement Current

6.1 On Conduction Current

The theorist tries to give physical laws a representation constructed using mathematical symbols; this representation should be as simple as possible. The distinct quantities that serve to signify the qualities regarded as first and irreducible must be as few as possible. Then, so that new facts are discovered, of which experience has determined the laws, the physicist must strive to express these laws by means of the signs already in use in the theory, to formulate them by means of the already-defined quantities. It is only when he recognizes the vanity of such an attempt, the impossibility of making the new laws fit in the old theories, that he decides to introduce into physics some rarely-used quantities, to fix the properties of these quantities by hypotheses that had not yet been uttered.

Thus, when Oersted, then Ampère, discovered and studied [108] electrodynamic and electromagnetic actions, physicists endeavored to make laws without introducing other quantities that had sufficed until then to represent all electrical and magnetic phenomena known in science: the *electric density* and *intensity of magnetization*. The exact knowledge of the distribution that affects the electricity spread on a conductor, at a given time, was, they thought, sufficient to determine the actions that this conductor has at this moment. Ampère did not believe these endeavors unworthy of his genius; but having finally recognized that they were condemned to ineffectualness, he imagined defining the properties of a wire at a given time by indicating not only what is, at this time and at each point of the wire, the value of the electric density, but also what the value of a new quantity, the *intensity of current* running through the wire, is.

If one takes the point of view of pure logic, the operation that involves introducing into a physical theory new quantities to represent new properties is entirely arbitrary; in fact, the theorist can be guided, in this operation, by loads of considerations extraneous to the field of physics, particularly by hypotheses that the philosophical doctrines on which he relies suggest regarding the nature of the phenomena studied,

© Springer International Publishing Switzerland 2015
P.M.M. Duhem, *The Electric Theories of J. Clerk Maxwell*,
Boston Studies in the Philosophy and History of Science 314,
DOI 10.1007/978-3-319-18515-6_6

explanations that are held in his time and country. Thus, to define the quantities to reduce theoretically the laws of attraction and electrical repulsion, physicists were inspired by the opinion which attributed these actions to a fluid or to two fluids. Similarly, to define quantities to represent electrodynamic phenomena, they allowed themselves to be guided by the idea that a *current* of an electric fluid running through the interpolar conductor, and they have imitated the formulas which, since Euler, were used to study the flow of a fluid.

The hydrodynamic analogy had already provided for Fourier the system of mathematical symbols by which he is able to represent the propagation of heat by conduction; it has provided for G.S. Ohm, Smaasen, and G. Kirchhoff the means for complementing, in the sense shown by Ampère, the mathematical representation of electrical phenomena. [109]

In imitation of the *speed* that, at each point, a flowing fluid offers, we imagine, at each point of the conductive body and at every moment, a directed quantity, the *electric current*.

Between the components of the speed of a moving fluid and the fluid density is a relationship, the *continuity relation*; in imitation of this relationship, one admits— among the components u, v, w of electrical current which relates to point (x, y, z) of the conductor at time t, and solid electrical density σ at the same point and at the same time—the existence of the equality

$$\frac{\partial u}{\partial x} + \frac{\partial v}{\partial y} + \frac{\partial w}{\partial z} + \frac{\partial \sigma}{\partial t} = 0. \tag{6.1}$$

To this relationship we add one concerning the surface density Σ at a point on the surface of two separate media 1 and 2:

$$u_1 \cos (N_1, x) + v_1 \cos (N_1, y) + w_1 \cos (N_1, z)$$
$$+ u_2 \cos (N_2, x) + v_2 \cos (N_2, y) + w_2 \cos (N_2, z) + \frac{\partial \Sigma}{\partial t} = 0. \tag{6.2}$$

In the spirit of the first physicists who considered them, the quantities u, v, w represented, at each point and at every moment, the components of the speed with which the electric fluid moves; we must not hesitate, today, to leave aside any hypothesis of this kind and simply regard u, v, w as three certain quantities varying with the coordinates and with the time and satisfying equalities (6.1) and (6.2).

To know completely the properties of a conductor at an *isolated instant* t, you need to know, at any point of the conductor, the values of the variables u, v, w, σ, and, furthermore, at any point of discontinuity of the surfaces, the value of the variable Σ. When it is proposed to set the properties of a conductor at all times for a certain period of time, it is only at the initial moment that the values of the five quantities σ, Σ, u, v, w should be given; at other times, it suffices to give the values of the variables w, v, w; σ, Σ are deduced by integrating Eqs. (6.1) and (6.2). [110]

6.2 On the Displacement Current

To represent the known laws that govern the actions of dielectric bodies, Faraday, Mossotti and their successors simply consider one directed quantity, varying from one point to another and from one moment to the other: the *intensity of polarization* of the components A, B, C.

Although no experience, at the time when he wrote, either justified or suggested even a similar hypothesis, Maxwell admitted that knowledge, at an *isolated instant t*, of the duration, of three components A, B, C, and of the polarization does not completely determine the properties of the dielectric at this instant. This body possessed properties, although unknown, which, at time t, depended not only on the intensity of polarization or *displacement*, but also on the *displacement current*, the directed quantity with components

$$\overline{u} = \frac{\partial A}{\partial t}, \quad \overline{v} = \frac{\partial B}{\partial t}, \quad \overline{w} = \frac{\partial C}{\partial t}. \tag{6.3}$$

The six variables A, B, C, $\overline{u}$, $\overline{v}$, $\overline{w}$ have values that can be chosen arbitrarily for an *isolated instant*, but it is not the same for all the moments of a certain period of time; if, for all these moments, we know the values of A, B, C, we know by the same fact the values of $\overline{u}$, $\overline{v}$, $\overline{w}$.

If we present it, as we have just done, as the purely arbitrary introduction of a new quantity of which no experience demands its employment, the definition, given by Maxwell, of the displacement current appears strange. On the contrary, it becomes very natural and, so to speak, forced, after taking into account the historical and psychological circumstances.

During his research on dielectrics, Maxwell, we have seen, constantly draws inspiration from the hypotheses of Faraday and Mossotti. In imitation of what Coulomb and Poisson had assumed for magnets, Faraday and Mossotti imagined a dielectric as a cluster of small conductive grains embedded in an insulating cement, each small conductive grain bearing [111] as much positive electricity as negative electricity; surely Maxwell, in all his writings, regards this image, if not as a depiction of reality, then at least as a model suggesting propositions that are always satisfied.

If, with Faraday and Mossotti, we regard a polarized dielectric as a set of conductive molecules on which electricity is distributed in a certain way, any change in the state of polarization of the dielectric consists in a change in the electrical distribution on the conductive molecules; this change in the polarization of the dielectric is therefore accompanied by real electric currents, each of which is located in a very small space. Moreover, we see without difficulty that these currents correspond, at each point of the dielectric, to an average current whose components are precisely given by equalities (6.3). This average current is therefore something other than the displacement current.

In his memoir: *On Physical Lines of Force*, Maxwell wrote,[1] by inviting his reader to refer to the work of Mossotti:

Electromotive force acting on a dielectric produces a state of polarization of its parts similar in distribution to the polarity of the particles of iron under the influence of a magnet, and, like the magnetic polarization, capable of being described as a state in which every particle has its poles in opposite conditions.

In a dielectric under induction, we may conceive that the electricity in each molecule is so displaced that one side is rendered positively, and the other negatively electrical, but that the electricity remains entirely connected with the molecule, and does not pass from one molecule to another.

The effect of this action on the whole dielectric mass is to produce a general displacement of the electricity in a certain direction. This displacement does not amount to a current, because when it has attained a certain value it remains constant, but it is the commencement of a current, and its variations constitute currents in the positive or [112] negative direction, according as the displacement is increasing or diminishing. The amount of the displacement depends on the nature of the body, and on the electromotive force; so that if h is the displacement, R the electromotive force, and E a coefficient[2] depending on the nature of the dielectric,

$$R = -4\pi E^2 h;$$

and if r is the value of the electric current due to displacement,

$$r = \frac{\partial h}{\partial t}.$$

This passage, the first where Maxwell has mentioned the displacement current, carries the indisputable mark of the ideas of Mossotti that led the Scottish physicist to imagine this current.

He explains so exactly, moreover, the conception that Maxwell formulated of this current, that we find it reproduced almost verbatim in the memoir: *A Dynamical Theory of the Electromagnetic Field*.[3] In the *Treatise on Electricity and Magnetism*,[4] we read this very brief passage:

The variations of electric displacement evidently constitute electric currents. These currents, however, can only exist during the variation of the displacement, and therefore, since the displacement cannot exceed a certain value without causing disruptive discharge, they cannot be continued indefinitely in the same direction, like the currents through conductors.

Maxwell adds[5]:

That whatever electricity may be, and whatever we may understand by the movement of electricity, the phenomenon which we have called electric displacement is a [113] movement of electricity in the same sense as the transference of a definite quantity of electricity through a wire is a movement of electricity....

---

[1]J. Clerk Maxwell, SCIENTIFIC PAPERS, vol. I, p. 491.

[2]We have insisted [1st Part, Chap. 4] on the sign error that affects this equality.

[3]J. Clerk Maxwell, *Scientific Papers*, vol. I, p. 531.

[4]J. Clerk Maxwell, *Treatise on Electricity and Magnetism*, trad. française, t. I, p. 69 [62].

[5]J. Clerk Maxwell, *Treatise on Electricity and Magnetism*, trad. française, t. I, p. 73 [66].

A displacement current is therefore essentially, and in the same way as a conduction current, a flow of electricity; in any conductive body, dielectric or magnetic, it produces the same induction, the same magnetization, and the same electrodynamic or electromagnetic forces as a conduction current of the same magnitude and direction. A current or a magnet exerts the same forces on a dielectric traversed by a displacement current which, on a conductor, would occupy the place as the dielectric and whose mass would be covered by a conduction current equal to the displacement current.

So we should never include, in electrodynamic calculations, the conduction current separately, whose components are u, v, w; still, will need to consider the *total current*, the geometric sum of the conduction current and the displacement current, of which $\bar{u}$, $\bar{v}$, $\bar{w}$ are the components. This principle is applied by Maxwell in his various writings on electricity[6]; it constitutes one of the foundations of his electrodynamic doctrine, one of its boldest and most productive innovations, as he himself remarks in this passage[7]:

> One of the chief peculiarities of this treatise is the doctrine which it asserts, that the true electric current $\mathfrak{C}$, that on which the electromagnetic phenomena depend, is not the same thing as $\mathfrak{K}$, the current of conduction, but that the time-variation of $\mathfrak{D}$, the electric displacement, must be taken into account in estimating the total movement of electricity....

[114]

6.3 In Maxwell's Theory, Is the Total Current a Uniform Current?

Suppose that at each point taken within a continuous domain we have the equality

$$\frac{\partial u}{\partial x} + \frac{\partial v}{\partial y} + \frac{\partial w}{\partial z} = 0, \tag{6.4}$$

and, at each point of a surface of discontinuity, that we have the equality

$$u_1 \cos (N_1, x) + v_1 \cos (N_1, y) + w_1 \cos (N_1, z)$$
$$+ u_2 \cos (N_2, x) + v_2 \cos (N_2, y) + w_2 \cos (N_2, z) = 0. \tag{6.5}$$

Then, we will have, at the first point, in virtue of equality (6.1),

$$\frac{\partial \sigma}{\partial t} = 0,$$

---

[6] J. Clerk Maxwell, *On Physical Lines of Force* (SCIENTIFIC PAPERS, vol. I, p. 496).—*A Dynamical Theory of the Electromagnetic Field* (SCIENTIFIC PAPERS, vol. I, p. 554).—TREATISE ON ELECTRICITY AND MAGNETISM, trad. française, t. II, p. 288 [232].

[7] J. Clerk Maxwell, *Treatise on Electricity and Magnetism*, trad. française, t. II, p. 288 [232–233].

and, at the second point, in virtue of equality (6.2),

$$\frac{\partial \Sigma}{\partial t} = 0.$$

The distribution of actual electricity on the system will remain invariable.

We give the name *uniform conduction current* to conduction currents which satisfy equalities (6.4) and (6.5).

Uniform displacement currents are displacement currents that satisfy equality

$$\frac{\partial \overline{u}}{\partial x} + \frac{\partial \overline{v}}{\partial y} + \frac{\partial \overline{w}}{\partial z} = 0 \tag{6.6}$$

at any point of a continuous medium and equality

$$\overline{u}_1 \cos (N_1, x) + \overline{v}_1 \cos (N_1, y) + \overline{w}_1 \cos (N_1, z)$$
$$+ \overline{u}_2 \cos (N_2, x) + \overline{v}_2 \cos (N_2, y) + \overline{w}_2 \cos (N_2, z) = 0 \tag{6.7}$$

at any point of a surface of discontinuity. [115]

If one accepts the definition of the fictitious electric densities e, E given by equalities (2.13) and (2.14) of the first part:

$$e = -\left(\frac{\partial A}{\partial x} + \frac{\partial B}{\partial y} + \frac{\partial C}{\partial z} \right), \tag{6.8}$$

$$E = -[A_1 \cos (n_1, x) + B_1 \cos (n_1, y) + C_1 \cos (n_1, z)$$
$$+ A_2 \cos (n_2, x) + B_2 \cos (n_2, y) + C_2 \cos (n_2, z)], \tag{6.9}$$

we can write, in general, in virtue of equalities (6.3),

$$\frac{\partial \overline{u}}{\partial x} + \frac{\partial \overline{v}}{\partial y} + \frac{\partial \overline{w}}{\partial z} + \frac{\partial e}{\partial t} = 0, \tag{6.10}$$

$$\overline{u}_1 \cos (N_1, x) + \overline{v}_1 \cos (N_1, y) + \overline{w}_1 \cos (N_1, z)$$
$$+ \overline{u}_2 \cos (N_2, x) + \overline{v}_2 \cos (N_2, y) + \overline{w}_2 \cos (N_2, z) + \frac{\partial E}{\partial t} = 0. \tag{6.11}$$

The uniform displacement current therefore satisfies equalities

$$\frac{\partial e}{\partial t} = 0, \quad \frac{\partial E}{\partial t} = 0;$$

hence, in any system, the invariability of the fictitious distribution of electricity equivalent to the dielectric polarization results.

It may happen that neither the conduction currents nor displacement currents are separately uniform, but that *the total current*, whose components are $(u+\overline{u})$, $(v+\overline{v})$, $(w+\overline{w})$, is uniform; it will satisfy, at any point of a continuous medium, the equality

$$\frac{\partial}{\partial x}(u+\overline{u}) + \frac{\partial}{\partial y}(v+\overline{v}) + \frac{\partial}{\partial z}(w+\overline{w}) = 0, \tag{6.12}$$

and, at any point of a surface of discontinuity, the equality

$$(u_1+\overline{u}_1)\cos(N_1,x) + (v_1+\overline{v}_1)\cos(N_1,y) + (w_1+\overline{w}_1)\cos(N_1,z)$$
$$+(u_2+\overline{u}_2)\cos(N_2,x) + (v_2+\overline{v}_2)\cos(N_2,y) + (w_2+\overline{w}_2)\cos(N_2,z) = 0. \tag{6.13}$$

[116] From these equalities (6.12) and (6.13), in virtue of equalities (6.1), (6.2), (6.10), and (6.11), the equalities

$$\frac{\partial}{\partial t}(\sigma+e) = 0, \tag{6.14}$$

$$\frac{\partial}{\partial t}(\Sigma+E) = 0 \tag{6.15}$$

result.

The actual electric distribution may vary from one moment to another; it is the same with the fictitious distribution equivalent to the dielectric polarization; but at each point of a continuous medium, or a surface of discontinuity, the sum of the actual electric density and fictitious electric density maintains a value independent of time, such that the electrostatic actions that are exerted in the system remain the same from one moment to the next moment.

To admit that the total current is always unchanging would be, for him who would at the same time recognize the legitimacy of all the previous equations, to deny the best-observed electrostatic phenomena; for example, it would be to deny that a capacitor can discharge through a stationary conductor placed between the two armatures.

The hypothesis that in any system, in any circumstances, the total current is always unchanging is, according to all the commentators of Maxwell, one of the essential principles underlying the doctrine of the Scottish physicist. Let us follow, in his writings, the formation of this hypothesis.

In the memoir: *On Faraday's Lines of Force*, the first that Maxwell devoted to the theories of electricity, there is no question yet regarding displacement current: the conduction current is only considered; what is said is easily consistent with the general considerations that we have outlined in Sect. 6.1. In particular, Maxwell admits[8] that the sum $\left(\dfrac{\partial u}{\partial x} + \dfrac{\partial v}{\partial y} + \dfrac{\partial w}{\partial z}\right)$ has a value, generally different from 0,

---

[8] J. Clerk Maxwell, SCIENTIFIC PAPERS, vol. I, p. 192.

[117] which he designates by $-4\pi\rho$. He adds only these words: "In a large class of phenomena, including all cases of uniform currents, the quantity ρ disappears.".

On the following page, based on the well-known electromagnetic properties of a closed[9] current,[10] Maxwell shows that the three components u, v, w of the conduction current can be put in the form

$$
\begin{cases}
-u = \dfrac{\partial \gamma}{\partial y} - \dfrac{\partial \beta}{\partial z}, \\[2mm]
-v = \dfrac{\partial \alpha}{\partial z} - \dfrac{\partial \gamma}{\partial x}, \\[2mm]
-w = \dfrac{\partial \beta}{\partial x} - \dfrac{\partial \alpha}{\partial y},
\end{cases}
\tag{6.16}
$$

α, β, γ being three functions of x, y, z which he calls the components of *the magnetic intensity*. From these equalities relationship (6.4) visibly follows; they therefore apply only to uniform currents. This conclusion ought not be surprising, the uniformity of the current being postulated in the same premises of the reasoning which gives equalities (6.16).

Maxwell notes this conclusion, but he did not care to deduce the impossibility of non-uniform currents. He said[11]:

We may observe that the above equations give by differentiation,

$$
\frac{\partial u}{\partial x} + \frac{\partial u}{\partial x} + \frac{\partial u}{\partial x} = 0,
$$

which is the equation of continuity for closed currents. Our investigations are therefore for the present limited to closed currents; and we know little of the magnetic effects of any currents which are not closed.

[118] The distinction between the conduction and displacement currents is introduced into the work of Maxwell in the memoir: *On Physical Lines of Force*.

At the point (x, y, z), the instantaneous rotational velocity of the ether has components α, β, γ; this speed represents,[12] in the kinetic theory that Maxwell develops in this memoir, the *intensity of the magnetic field*; then posing

$$
\begin{cases}
\dfrac{\partial \gamma}{\partial y} - \dfrac{\partial \beta}{\partial z} - 4\pi u, \\[2mm]
\dfrac{\partial \alpha}{\partial z} - \dfrac{\partial \gamma}{\partial x} - 4\pi v, \\[2mm]
\dfrac{\partial \beta}{\partial x} - \dfrac{\partial \alpha}{\partial y} - 4\pi w,
\end{cases}
\tag{6.17}
$$

---

[9][in the sense of "closed circuit"].

[10]We will return to this demonstration in Chap. 7, Sect. 7.1.

[11]J. Clerk Maxwell, *loc. cit.*, p. 195.

[12]J. Clerk Maxwell, SCIENTIFIC PAPERS, vol. I, p. 460.

Maxwell admits[13] that u, v, w represent, at the point (x, y, z), the components of the conduction current; *the conduction current is therefore uniform by definition.* This proposition thus has nothing surprising in a writing where, implicitly, the true electric density σ is always assumed to be equal to 0 and where only the fictitious electric density e, equivalent to the dielectric polarization, is introduced.

It is related[14] to the components of the total current by the continuity equation:

$$\frac{\partial}{\partial x}(u + \overline{u}) + \frac{\partial}{\partial y}(v + \overline{v}) + \frac{\partial}{\partial z}(w + \overline{w}) + \frac{\partial e}{\partial t} = 0, \tag{6.18}$$

which can also be written, because of equality (6.17),

$$\frac{\partial \overline{u}}{\partial x} + \frac{\partial \overline{v}}{\partial y} + \frac{\partial \overline{w}}{\partial z} + \frac{\partial e}{\partial t} = 0.$$

If, therefore, in this memoir Maxwell defines conduction current [119] as being essentially uniform, he only poses the same hypothesis regarding displacement currents.

It is similar in the memoir: *A Dynamical Theory of the Electromagnetic Field.* Using the known laws of electromagnetism, laws which essentially involve closed and uniform currents, Maxwell establishes[15] Eq. (6.17), which he considers as applying to all conduction currents; so he admits there that these currents are always uniform. But he is careful to extend this proposition to the total current; this satisfies[16] equality (6.18), resulting, for the displacement current, in equality (6.19).

When, in this same memoir, Maxwell develops the theory of the propagation of displacement current in a dielectric medium, he is careful to claim that these currents are always and necessarily transverse, subject to the condition

$$\frac{\partial \overline{u}}{\partial x} + \frac{\partial \overline{v}}{\partial y} + \frac{\partial \overline{w}}{\partial z} = 0.$$

He admits, instead, that at each point the apparent density may vary from one moment to the next, and he establishes[17] the law which governs this variation; however, to get rid of the longitudinal current which would thus be introduced and which would hinder the electromagnetic theory of light, he adds these words: "Since the medium is a perfect insulator, e, the free electricity, is immoveable..." Nothing in the ideas put forward by Maxwell in the course of this memoir or his previous writings justified this conclusion; the density e, linked to variations in the electric displacement, depends in no way on the conduction current.

---

[13] J. Clerk Maxwell, *loc. cit.*, vol. I, p. 462.

[14] J. Clerk Maxwell, *loc. cit.*, equality (113), vol. I, p. 496.

[15] J. Clerk Maxwell, SCIENTIFIC PAPERS, vol. I, p. 557.

[16] J. Clerk Maxwell, *loc. cit.*, p. 561, equality (H).

[17] J. Clerk Maxwell, *loc. cit.*, vol. I, p. 582.

The electromagnetic theory of light, however, requires that the displacement current in a non-conducting dielectric propagates according to the same laws as small movements in an elastic and non-compressible solid; the principles laid down by Maxwell in his memoirs do not meet this [120] requirement. It is not the same for the singular theory that Maxwell develops in his *Treatise on Electricity and Magnetism* and that we named his *third electrostatics*.

There is electrical charge nowhere else than the fictitious charge due to the dielectric polarization, no density than the densities e, E; it is to these densities that the components of the conduction current will be linked by continuity relations taken in their usual form. At any point of a continuous medium, we have[18]

$$\frac{\partial u}{\partial x} + \frac{\partial v}{\partial y} + \frac{\partial w}{\partial z} + \frac{\partial e}{\partial t} = 0. \tag{6.19}$$

At any point of a surface of discontinuity, we have[19]

$$u_1 \cos{(N_1, x)} + v_1 \cos{(N_1, y)} + w_1 \cos{(N_1, z)}$$
$$+ u_2 \cos{(N_2, x)} + v_2 \cos{(N_2, y)} + w_2 \cos{(N_2, z)} + \frac{\partial E}{\partial t} = 0. \tag{6.20}$$

But, on the other hand, densities e, E are related to components A, B, C of the intensity of the dielectric polarization, which Maxwell designates by f, g, h and calls the components of *displacement*. The relationship between these quantities is given by the following equalities, which we discussed in the first part of this work[20] and which Maxwell is careful to recall[21] with the equalities that we just wrote:

$$e = \frac{\partial A}{\partial x} + \frac{\partial B}{\partial y} + \frac{\partial C}{\partial z}, \tag{6.21}$$

$$E = A_1 \cos{(N_1, x)} + B_1 \cos{(N_1, y)} + C_1 \cos{(N_1, z)}$$
$$+ A_2 \cos{(N_2, x)} + B_2 \cos{(N_2, y)} + C_2 \cos{(N_2, z)}. \tag{6.22}$$

[121] Differentiating these equalities with respect to t, and taking account of equalities (6.3) that define displacement currents, we find

$$\frac{\partial \overline{u}}{\partial x} + \frac{\partial \overline{v}}{\partial y} + \frac{\partial \overline{w}}{\partial z} - \frac{\partial \overline{e}}{\partial t} = 0, \tag{6.23}$$

---

[18] J. Clerk Maxwell, *Treatise on Electricity and Magnetism*, trad. française, t. I, p. 506 [412], equality (2). It should be noted that this passage contradicts what gives Maxwell on p. 470 [380], where he seems to admit that any conduction current is uniform, in accordance with his old ideas.

[19] J. Clerk Maxwell, *loc. cit.*, t. I, p. 510 [415], equality (5).

[20] 1st Part, equalities (5.4) and (5.5).

[21] J. Clerk Maxwell, *Treatise on Electricity and Magnetism*, trad. française, t. 1, p. 506 [412], equality (1) and p. 510 [415], equality (4).

$$\overline{u}_1 \cos{(N_1, x)} + \overline{v}_1 \cos{(N_1, y)} + \overline{w}_1 \cos{(N_1, z)}$$

$$+ \overline{u}_2 \cos{(N_2, x)} + \overline{v}_2 \cos{(N_2, y)} + \overline{w}_2 \cos{(N_2, z)} - \frac{\partial E}{\partial t} = 0. \qquad (6.24)$$

As was to be expected, these equalities differ by the sign of the terms in $\dfrac{\partial e}{\partial t}$, $\dfrac{\partial E}{\partial t}$, of equalities (6.10) and (6.11), which stem from the usual theory of dielectric polarization that Maxwell admitted before having conceived the special electrostatics that was developed in his *Treatise*.

Adding the terms of equalities (6.19) and (6.23), on the one hand, and equalities (6.20) and (6.24), on the other hand, we find at any point of a continuous medium the equality

$$\frac{\partial}{\partial x}(u + \overline{u}) + \frac{\partial}{\partial y}(v + \overline{v}) + \frac{\partial}{\partial z}(w + \overline{w}) = 0, \qquad (6.25)$$

and, at any point of a surface of discontinuity, equality

$$(u_1 + \overline{u}_1) \cos{(N_1, x)} + (v_1 + \overline{v}_1) \cos{(N_1, y)} + (w_1 + \overline{w}_1) \cos{(N_1, z)}$$

$$+ (u_2 + \overline{u}_2) \cos{(N_2, x)} + (v_2 + \overline{v}_2) \cos{(N_2, y)} + (w_2 + \overline{w}_2) \cos{(N_2, z)} = 0. \qquad (6.26)$$

So, therefore, the latest electrostatic theory adopted by Maxwell leads to the following consequences:

Not only within a continuous medium do the components of the total current satisfy the same relationship as the components of the current within an incompressible fluid, but also, on the surface of separation of two different media, the total current experiences no abrupt change, neither in magnitude nor direction. The total current in any system corresponds to a closed and uniform current.

Since the moment Maxwell conceived his third electrostatics, he saw this consequence, so favorable to his ideas on the electromagnetic theory [122] of light. In a note,[22] where he remarked that the polarization of a dielectric plate placed between two conductors is directed from conductor A, positively electrified, to conductor B, negatively electrified, noting that it led him necessarily to his third electrostatics, since he admitted no other electrification than the fictitious electrification, he added:

> Thus, if the two conductors in the last case are now joined by a wire, there will be a current in the wire from A to B. At the same time, since the electric displacement in the dielectric is diminishing, there will be an action electromagnetically equivalent to that of an electric current from B to A through the dielectric. According to this view, the current produced in discharging a condenser is a complete circuit...

---

[22] J. Clerk Maxwell, *On a Method of Making a Direct Comparison of Electrostatic with Electromagnetic Force: With a Note on the Electromagnetic Theory of Light*, read at the Royal Society of London on 18 June 1868 (PHILOSOPHICAL TRANSACTIONS, vol. CLVIII.—SCIENTIFIC PAPERS, vol. II, p. 139).

Later, in his *Treatise on Electricity and Magnetism*, Maxwell takes up[23] the same considerations with more developments. He says:

> ...let us consider an accumulator formed of two conducting plates A and B, separated by a stratum of a dielectric C. Let W be a conducting wire joining A and B, and let us suppose that by the action of an electromotive force a quantity Q of positive electricity is transferred along the wire from B to A. ...at the same time that a quantity Q of electricity is being transferred along the wire by the electromotive force from B towards A, so as to cross every section of the wire, the same quantity of electricity crosses every section of the dielectric from A towards B by reason of the electric displacement.
>
> The displacements of electricity during the discharge of the accumulator will be the reverse of these. In the wire the discharge will be Q from A to B, and in the dielectric the displacement will subside, and a quantity of electricity Q will cross every section from B towards A.
>
> Every case of charge or discharge may therefore be considered as a motion in a closed circuit, [123] such that at every section of the circuit the same quantity of electricity crosses in the same time, and this is the case, not only in the voltaic circuit where it has always been recognised, but in those cases in which electricity has been generally supposed to be accumulated in certain places.
>
> We are thus led to a very remarkable consequence of the theory which we are examining, namely, that the motions of electricity are like those of an *incompressible* fluid...

This passage is followed, in the *Treatise* of Maxwell, by the following sentence: "...Thus when the charged conductor is introduced into the closed space there is immediately a displacement of a quantity of electricity equal to the charge."

By writing this sentence, Maxwell forgets, for a moment, the very special meaning which this proposition has in his latest theory: *the total current is uniform*, to restore the meaning that it has in the minds of most physicists, which it had in his early writings. But this is an apparent oversight. It is quite true that the components of the total current satisfy relations (6.25) and (6.26), similar to those that characterize a uniform current; but it is not true that the amount of electricity within a given space is either always invariable, nor that the quantities $\dfrac{\partial e}{\partial t}$, $\dfrac{\partial E}{\partial t}$ are all equal to 0. It is one of the paradoxical characters of Maxwell's latest theory that the uniformity of the total current does not entail the invariability of the electrical distribution nor of electrostatic actions.

However, if it is not true that the amount of electricity in a closed surface always remains invariable, this proposition being true when the closed surface contains only displacement currents, without a trace of conduction current—or even that conduction current, without a trace of displacement current; it suffices, to convince oneself, to glance either at equalities (6.19) and (6.20), or at equalities (6.21) and (6.22). So when Maxwell, developing in his treatise the electromagnetic theory of light, writes[24]: "If the medium is a non-conductor, ...the volume-density [124] of free electricity, is independent of t," he affirms a necessary consequence of the doctrine developed in this *Treatise*. Whereas in the same sentence written by him on the same

---

[23] J. Clerk Maxwell, *Treatise on Electricity and Magnetism*, trad. française, t. I, p. 71 [63–64].

[24] J. Clerk Maxwell, *Treatise on Electricity and Magnetism*, trad. française, t. II, p. 488 [385].

occasion in his memoir: *A Dynamical Theory of the Electromagnetic Field*, there
was a fallacy, in contradiction with the ideas accepted in this memoir.

But if the electrical distribution can vary within a conductive non-dielectric body,
no more than within a non-conducting dielectric, this distribution can vary from
one moment to the next at the surface through which a conducting medium borders a
dielectric medium. These variations give rise to the charge and discharge phenomena
that are studied in electrostatics.

6.4   Return to the Third Maxwell Electrostatics. To What Extent it Can Agree with Classical Electrostatics

Maxwell, we saw [1st part, Chap. 5, Sect. 5.3] avoids, in his third electrostatics,
establishing between the function Ψ and densities e, E only equalities (5.16) and
(5.18); therefore, one would believe that it is permissible to repeat here what we
said in the 1st part, Chap. 4, Sect. 4.4, denouncing as illusory the third electrostatics
of Maxwell, declaring that it does not contain the necessary elements to cast into
equations the least problem of electrical distribution.

One is all the more tempted to formulate a similar judgment that, in his *Treatise on
Electricity and Magnetism*, Maxwell makes no use of this electrostatics; he does even
adopt the very solution of the two problems that, in his memoirs: *On Physical Lines
of Force* and *A Dynamical Theory of the Electromagnetic Field*, he had attempted
to resolve. He treats neither the theory of capacitors nor the theory of forces exerted
between electrified bodies.

No doubt, in his *Treatise* we read chapters or portions of chapters dealing with
electrical distributions or electrostatic forces. But the reasoning that he develops, the
formulas that are used, are in no way particular to the electrostatics whose principles
we analyzed; [125] they both depend on the electrostatics founded on Coulomb's
laws, on the classical electrostatics created by Poisson.

However, the judgment we just outlined would be unfair; we can, in the system
of Maxwell, obtain a casting into equations of the electrostatic problem. It suffices
to introduce suitable hypotheses that will replace the analytical expression of the
potential function deduced, in the ordinary theory, from Coulomb's laws.

And first, inside a conductive body, the components of the conduction current are
proportional to the components of the electromotive force. So there is equilibrium,
it is necessary that the first vanish and, therefore the second, which the equalities

$$\frac{\partial \Psi}{\partial x} = 0, \quad \frac{\partial \Psi}{\partial y} = 0, \quad \frac{\partial \Psi}{\partial z} = 0$$

express. On a same conductive mass, the function Ψ will have, at any point, the
same value.

Inside of a non-conductive body, the conduction current is everywhere zero. Therefore, equalities (6.25) and (6.26) become

$$\frac{\partial \overline{u}}{\partial x} + \frac{\partial \overline{v}}{\partial y} + \frac{\partial \overline{w}}{\partial z} - \frac{\partial \overline{e}}{\partial t} = 0,$$

$$\overline{u}_1 \cos (N_1, x) + \overline{v}_1 \cos (N_1, y) + \overline{w}_1 \cos (N_1, z)$$
$$+ \overline{u}_2 \cos (N_2, x) + \overline{v}_2 \cos (N_2, y) + \overline{w}_2 \cos (N_2, z) = 0$$

or else, in virtue of equalities (6.23) and (6.24),

$$\frac{\partial e}{\partial t} = 0, \quad \frac{\partial E}{\partial t} = 0.$$

Inside a continuous insulating body or on the surface of contact between two different insulating bodies, the distribution of electricity is invariable. It will be assumed, in general, that the two densities are equal to 0:

$$e = 0, \quad E = 0.$$

[126] Indeed, this hypothesis is not explicitly stated in the writings of Maxwell, but we can say that it is there implicitly; every moment, Maxwell, we have seen, repeats that the electric charge, a residual effect of polarization, does not feel outside of the dielectric, but only on the surface of contact with the conductor and the dielectric; also, we have quoted passages of Faraday and Mossotti where these authors expressed a similar opinion. We will therefore interpret the thought of Maxwell without bias by expressing that both of the electrical densities are zero in any insulating medium.

In the classical theory, it should be noticed, we are compelled to introduce an hypothesis which has analogies with the previous one; there, alongside the dielectric polarization and the *fictional* electric charge that is equivalent to it, we considered a *true* electric charge. On a non-conductive body, the latter creates an invariant distribution that, in each problem, must be regarded as given; and, in most cases, one assumes that the true electric charge is zero at any point of the insulating bodies considered; but this hypothesis does not prejudge anything about the fictitious electrification and polarization to which it is equivalent.

In the system of Maxwell, we do not encounter real electric charge next to the apparent electrical charge which is equivalent to the dielectric polarization; only the latter exists. To it belongs, on poorly conducting bodies, the character of invariability, attributed by the classical theory to true electric charge; this is what must be regarded as a given.

If the densities e, E equal zero, equalities (5.16) and (5.17) of the first part are transformed into equality

$$\frac{\partial}{\partial x}\left(K\frac{\partial \Psi}{\partial x}\right) + \frac{\partial}{\partial y}\left(K\frac{\partial \Psi}{\partial y}\right) + \frac{\partial}{\partial z}\left(K\frac{\partial \Psi}{\partial z}\right) = 0,$$

satisfied at any point of a continuous insulating medium, and into equality

$$K_1\frac{\partial \Psi}{\partial N_1} + K_2\frac{\partial \Psi}{\partial N_2} = 0,$$

[127] satisfied at the surface of separation of two separate insulating media.

One thus obtains the equations to determine the function Ψ; and, most importantly, these equations are what would be used to determine the electrostatic potential function, according to the classical theory, in a system where each dielectric would have a specific inductive capacity proportional to K.

The analogy between Maxwell's theory and classical theory is complete, where conductors are immersed in *a single homogeneous dielectric*. In this case, the function Ψ, constant within each conductor, must satisfy in the interposed space equality $\Delta\Psi = 0$; once determined by these conditions, the function Ψ in turn determines the surface density on the surface of each conductor by equality (5.16) of the first part, which becomes

$$\frac{\partial \Psi}{\partial N_e} = -\frac{4\pi}{K}E.$$

It is clear, therefore, that one can write

$$\Psi = \frac{1}{K}\int \frac{E}{r}dS,$$

the integral extending over all electrified surfaces. The electrostatic energy then has the value

$$U = \frac{1}{2}\int \Psi E\, dS$$

or

$$\frac{1}{2K}\iint \frac{EE'}{r}\, dS\, dS'.$$

We compare these formulas with those that would give the classical theories, whose principles are recalled in the Chap. 2 of the first part.

Suppose that, in an non-polarizable medium, two electric charges q and q' separated by the distance r repel each other with a force $\varepsilon\frac{qq'}{r^2}$. We denote by F the coefficient of polarization of the [128] dielectric medium and by V the total electrostatic potential function that the sum $(V + \overline{V})$, in the indicated chapter, signifies. Let

Σ be the actual surface density of electricity; it corresponds to a total, both real and fictitious, density

$$\Delta = \frac{\Sigma}{1 + 4\pi\varepsilon F}.$$

Function V, constant on each conductive body, is harmonic in the dielectric; it is obviously the same for the function $\frac{\varepsilon V}{1+4\pi\varepsilon F}$.

We then have, in the area of contact of a conductor and the dielectric,

$$\frac{\partial V}{\partial N_e} = -4\pi\Delta,$$

which can be written

$$\frac{\partial}{\partial N_e}\frac{\varepsilon V}{1 + 4\pi\varepsilon F} = -\frac{4\pi\varepsilon}{(1 + 4\pi\varepsilon F)^2}\Sigma.$$

Finally, the electrostatic energy is set to

$$U = \frac{\varepsilon}{2}\iint \frac{\Delta\Delta'}{r}dS\,dS',$$

which can be written

$$U = \frac{\varepsilon}{2(1 + 4\pi\varepsilon F)}^2\iint \frac{\Sigma\Sigma'}{r}dS\,dS'.$$

We see that we pass from the formulas of Maxwell to these ones if we replace

$$E \quad \text{by} \quad \Sigma,$$
$$\Psi \quad " \quad \frac{\varepsilon V}{1 + 4\pi\varepsilon F},$$
$$\Psi \quad " \quad \frac{(1 + 4\pi\varepsilon F)^2}{\varepsilon}.$$

The analogy of the two theories is now complete. [129]

The analogy between Maxwell's theory and the classical theory is also complete where the system contains a heterogeneous dielectric or several separate dielectrics.

Suppose that conductors 1 are immersed in a uniform and undefined dielectric medium 0, and that, in this medium, lies another dielectric and homogeneous body 2; to the dielectrics 0 and 2 correspond the values K_0, K_2 of the coefficient K.

The function Ψ, which is continuous throughout space and constant within each of the conductors, satisfies the equation $\Delta\Psi = 0$ inside both of the dielectrics 0 and 2.

At the surface of separation between dielectric 0 and dielectric 2, it satisfies the relationship

$$K_0 \frac{\partial \Psi}{\partial N_0} + K_1 \frac{\partial \Psi}{\partial N_1} = 0. \tag{a}$$

At the surface of contact between body 1 and dielectric 0 is a surface density E_{10} given by the equality

$$E_{10} = -\frac{K_0}{4\pi} \frac{\partial \Psi}{\partial N_0}. \tag{b}$$

Finally, the electrostatic energy has the value

$$U = \frac{1}{2} \int \Psi E_{10} d S_{10}. \tag{c}$$

Compare these relationships with those given by the classical theory.

Function V, continuous throughout all space and constant inside conductors, is harmonic in the dielectric.

At the surface of dielectric 0 and 2, we have

$$(1 + 4\pi \varepsilon F_0) \frac{\partial V}{\partial N_0} + (1 + 4\pi \varepsilon F_2) \frac{\partial V}{\partial N_2} = 0. \tag{α}$$

[130] At the the the surface of contact of conductor 1 and the dielectric surface lies a real surface density

$$\Sigma_{10} = -\frac{1 + 4\pi \varepsilon F_0}{4\pi} \frac{\partial V}{\partial N_0}. \tag{β}$$

At the surface of contact between two dielectrics one finds a purely fictitious surface density

$$\Delta_{20} = -\frac{1}{4\pi} \left(\frac{\partial V}{\partial N_0} + \frac{\partial V}{\partial N_2} \right)$$
$$= -\frac{\varepsilon(F_2 - F_0)}{1 + 4\pi \varepsilon F_2} \frac{\partial V}{\partial N_0}, \tag{β'}$$

and this density is non-zero, in general, if F_2 is not equal to F_0.

Finally, the electrostatic energy has the value

$$U = \frac{\varepsilon}{2} \int V \Delta_{10} d S_{10} + \frac{\varepsilon}{2} \int V \Delta_{20} d S_{20}$$

or

$$U = \frac{\varepsilon}{2(1 + 4\pi \varepsilon F_0)} \int V \Sigma_{10} d S_{10} + \frac{\varepsilon}{2} \int V \Delta_{20} d S_{20}. \tag{γ}$$

Can we move from the first group of formulas to the second by replacing E_{10} by Σ_{10} and Ψ by λV, λ being a suitably chosen constant?

Comparison of equalities (b) and (β) would give

$$\frac{1 + 4\pi \varepsilon F_2}{1 + 4\pi \varepsilon F_0} = \frac{K_2}{K_0}.$$

We therefore have, in a general way,

$$K\lambda = (1 + 4\pi \varepsilon F).$$

[131] Equality (c) would become

$$U = \frac{\lambda}{2} \int V \Sigma_{10} \, dS_{10}.$$

If we put

$$\lambda = \frac{\varepsilon}{1 + 4\pi \varepsilon F_0},$$

we would find the first term of the expression (γ), but not the second.

So we come to the following conclusion:

If 0 refers to the ethereal polarizable medium where all bodies are supposed to be immersed; if F_0 is the coefficient of dielectric polarization of the medium; if F_2 is the coefficient of dielectric polarization of the body immersed in this medium; if, finally, in the equations of the third electrostatics of Maxwell one replaces:

The electric density E on the surface of conductors	by the actual electric density Σ,
The function Ψ	by the function $\frac{\varepsilon V}{1 + 4\pi \varepsilon F_0}$, where V is the electrostatic potential function,
The coefficient K_0	by $\frac{(1 + 4\pi \varepsilon F_0)^2}{\varepsilon}$,
The coefficient K_2	by $\frac{(1 + 4\pi \varepsilon F_0)(1 + 4\pi \varepsilon F_2)}{\varepsilon}$,
Therefore, the ratio $\frac{K_2}{K_0}$	by $\frac{1 + 4\pi \varepsilon F_2}{1 + 4\pi \varepsilon F_0}$,

we find the formulas by which classical electrostatics determines the value of the potential function in the entire system and the actual distribution of electricity on conductors, so that, for these problems, the two electrostatics provide equivalent solutions.

The equivalence continues if one wants to study ponderomotive forces produced between electrified conductors in a system that does not contain the other dielectric medium 0. [132]

But if there is another dielectric 2, the previous transformation applied to the electrostatic energy of Maxwell does not give the classical electrostatic energy. It lacks the term

$$\frac{\varepsilon}{2} \int V \Delta_{20} \, dS_{20} = -\frac{\varepsilon}{8\pi} \int V \left(\frac{\partial V}{\partial N_0} + \frac{\partial V}{\partial N_2} \right) dS_{20}$$

$$= -\frac{\varepsilon(F_2 - F_0)}{1 + 4\pi \varepsilon F_2} \int V \frac{\partial V}{\partial N_0} dS_{20},$$

which would also be written, in virtue of the above equivalences that we indicated,

$$-\frac{\varepsilon}{8\pi \lambda^2} \int \Psi \left(\frac{\partial \Psi}{\partial N_0} + \frac{\partial \Psi}{\partial N_2} \right) dS_{20} = -\frac{(1 + 4\pi \varepsilon F_0)^2)}{8\pi \varepsilon} \int \Psi \left(\frac{\partial \Psi}{\partial N_0} + \frac{\partial \Psi}{\partial N_2} \right) dS_{20}$$

$$= -\frac{K_0}{8\pi} \int \Psi \left(\frac{\partial \Psi}{\partial N_0} + \frac{\partial \Psi}{\partial N_2} \right) dS_{20}$$

$$= -\frac{K_2 - K_0}{8\pi} \int \Psi \frac{\partial \Psi}{\partial N_2} dS_{20}$$

$$= \frac{K_2 - K_0}{8\pi} \int_2 \left[\left(\frac{\partial \Psi}{\partial x} \right)^2 + \left(\frac{\partial \Psi}{\partial y} \right)^2 + \left(\frac{\partial \Psi}{\partial z} \right)^2 \right] d\omega_2.$$

We see that this term can be null if the electric field is not zero and dielectric 2 differs from the medium 0.

The presence or absence of this term will differentiate the law of ponderomotive forces that are exerted in the system according to the traditional doctrine or the doctrine of Maxwell.

However, the researches of Gouy[25] which are also on this point a natural sequel of ours,[26] showed that the classical doctrine was fully aware of the actions observed between conductors and dielectrics by various physicists, notably by Pellat. It must be concluded that in general these actions do not agree with the electrostatics of Maxwell. [133]

---

[25]Gouy, JOURNAL DE PHYSIQUE, 3° série, t. V, p. 154, 1896.
[26]P. Duhem, *Leçons sur l'Électricité et le Magnétisme*, t. II, 1892.

Chapter 7
The Six Equations of Maxwell and Electromagnetic Energy

7.1 The Three Relations Between the Components of the Electric Field and the Components of the Current

Suppose that a uniform electrical current flows through a wire disposed on the contour C of an area A; we look at this area in such a way as to see current circulate in a counterclockwise direction; we will look at the positive side of area A (Fig. 7.1).

If a magnetic pole, containing a unit of southern magnetism, is placed in the presence of this current, it is subjected to a force whose components are α, β, γ; this is what Maxwell called the *magnetic force*, what, more precisely, is today called the *magnetic field*.

Suppose that this unit pole describes a closed curve e, that this curve pierces once and only once the area A, and that it pierces it [134] from the negative side to the positive side; the force to which the pole is subject performs some work, which the integral

$$\int_c (\alpha \, dx + \beta \, dy + \gamma, dz),$$

extended to the closed curve c, represents.

The laws of electromagnetism, established by Biot, Savart, Laplace, Ampère, and Savary, make the properties of the magnitudes α, β, γ known. These laws lead to the following result:

The work of which we just gave the expression depends neither on the shape of the curve c nor on the shape of the curve C. It depends only on the current that runs through the curve C; if this intensity J is measured in electromagnetic units, it is $4\pi J$:

$$\int_c (\alpha \, dx + \beta \, dy + \gamma \, dz) = 4\pi J. \tag{7.1}$$

© Springer International Publishing Switzerland 2015
P.M.M. Duhem, *The Electric Theories of J. Clerk Maxwell*,
Boston Studies in the Philosophy and History of Science 314,
DOI 10.1007/978-3-319-18515-6_7

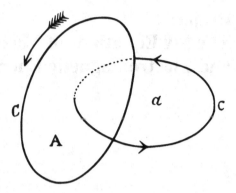

This equality can be understood somewhat differently. Suppose that the curve c is the contour of an area a. If we consider the area a such that we see the pole of the magnet turn in an counterclockwise direction, we will say that we look at the *positive side* of the area a.

It is clear that the current that runs through the wire G pierces the area a passing from the negative side to the positive side; and as, through each section of the wire G, it carries in the time dt, a quantity $dQ = J\,dt$ of positive electricity, we can say that the area a is crossed, during the time dt, from the negative side to the positive side, by a quantity of positive electricity $dQ = J\,dt$. So equality (7.1) can be written

$$dt \int_c (\alpha\,dx + \beta\,dy + \gamma\,dz) = 4\pi\,dQ. \qquad (7.2)$$

This equality easily extends to the case where the field contains any number of wires with closed and uniform currents. If a closed curve c, traveling in a determined direction, [135] is the outline of an area a and if dQ is the amount of positive electricity which, in time dt, pierces the area a from the negative side to the positive side, equality (7.2) remains exact.

The demonstration assumes that the curve c has no point in common with the wires that carry electricity; to free equality (7.2) from this limitation, some precautions may be necessary. Without delaying, Maxwell admits that equality (7.2) applies even in the case where the closed curve c is drawn within a body whose electrical currents flow continuously.

In this latter case, the amount dQ is simply related to these currents.

Let $d\sigma$ be an element of area a; u, v, w the components of the electric current at this point; N the normal to this element, oriented in the sense that it pierces the area a from the negative side to the positive side. In the same direction, and during the time dt, area $d\sigma$ gives passage to a quantity of electricity

$$[u \cos(N, x) + v \cos(N, y) + w \cos(N, z)]d\sigma\,dt$$

and the whole area a to a quantity of electricity

$$dt \int_a [u \cos(N, x) + v \cos(N, y) + w \cos(N, z)] d\sigma = dQ.$$

Equality (7.2) becomes

$$\int (\alpha \, dx + \beta \, dy + \gamma \, dz) - 4\pi \int_a [u \cos(N, x) + v \cos(N, y) + w \cos(N, z)] d\sigma = 0.$$
$$(7.3)$$

However, a formula often used by Ampère, and whose general form is due to Stokes, allows us to write

$$\int (\alpha \, dx + \beta \, dy + \gamma \, dz) = - \int_a \left[\left(\frac{\partial \gamma}{\partial y} - \frac{\partial \beta}{\partial z} \right) \cos(N, x) \right. $$
$$+ \left(\frac{\partial \alpha}{\partial z} - \frac{\partial \gamma}{\partial x} \right) \cos(N, y)$$
$$\left. + \left(\frac{\partial \beta}{\partial x} - \frac{\partial \alpha}{\partial y} \right) \cos(N, z) \right] d\sigma.$$

[136] Equality (7.3) can be written

$$\int_a \left[\left(\frac{\partial \gamma}{\partial y} - \frac{\partial \beta}{\partial z} + 4\pi u \right) \cos(N, x) + \left(\frac{\partial \alpha}{\partial z} - \frac{\partial \gamma}{\partial x} + 4\pi v \right) \cos(N, y) \right.$$
$$\left. + \left(\frac{\partial \beta}{\partial x} - \frac{\partial \alpha}{\partial y} + 4\pi w \right) \cos(N, z) \right] d\sigma = 0.$$

This equality must be true for any area a drawn inside the body in which the electrical currents flow. For this, one easily sees, it is necessary and right that there be, at any point in this body, three equalities

$$\begin{cases} \dfrac{\partial \gamma}{\partial y} - \dfrac{\partial \beta}{\partial z} = -4\pi u, \\[2mm] \dfrac{\partial \alpha}{\partial z} - \dfrac{\partial \gamma}{\partial x} = -4\pi v, \\[2mm] \dfrac{\partial \beta}{\partial x} - \dfrac{\partial \alpha}{\partial y} = -4\pi w. \end{cases} \qquad (7.4)$$

These three equations, to which Maxwell assigns a key role, are established, in his oldest memoir[1] on electricity, by a demonstration differing from the previous

---

[1] J. Clerk Maxwell, *On Faraday's Lines of Force* (SCIENTIFIC PAPERS, vol. I, p. 194). Actually, in this memoir, Maxwell omits the factor 4π; further, the signs of the second members are changed as a result of a different orientation of the coordinate axes.

one by simple nuances; he reproduced this demonstration[2] or sketched it[3] in all his subsequent writings.

In his memoir: *On Faraday's Lines of Force*,[4] Maxwell followed Eq. (7.4) with the remark:

We may observe that the above equations give by differentiation

$$\frac{\partial u}{\partial x} + \frac{\partial v}{\partial y} + \frac{\partial w}{\partial z} = 0,$$

[137] which is the equation of continuity for closed currents. Our investigations are therefore for the present limited to closed currents; and we know little of the magnetic effects of any currents which are not closed.

The condition of uniformity, imposed on currents in the premises of the reasoning, is reflected in the consequences. Maxwell, who, at the time he wrote the previous lines, professed, on electrical currents, the same ideas as all physicists, refrains from concluding that all the currents are necessarily uniform, but only the application of Eq. (7.4) is limited to uniform currents.

The same observation is found in the *Treatise on Electricity and Magnetism*; but, according to the doctrine set out in this *Treatise*, if the conduction and displacement currents can be separately non-uniform, the total flux, obtained by the composition of the previous two, is always uniform; Eq. (7.4) will therefore be exempt from any exception "if we take u, v, w as the components of that electric flow which is due to the variation of electric displacement as well as to true conduction."[5] In other words, we can, in any event, write the relations

$$\begin{cases} \dfrac{\partial \gamma}{\partial y} - \dfrac{\partial \beta}{\partial z} = -4\pi(u + \overline{u}), \\[2ex] \dfrac{\partial \alpha}{\partial z} - \dfrac{\partial \gamma}{\partial x} = -4\pi(v + \overline{v}), \\[2ex] \dfrac{\partial \beta}{\partial x} - \dfrac{\partial \alpha}{\partial y} = -4\pi(w + \overline{w}). \end{cases} \tag{7.5}$$

[2] J. Clerk Maxwell, *On Physical Lines of Force* (SCIENTIFIC PAPERS, vol. I, p. 462).—*Treatise on Electricity and Magnetism*, trad. française, t. II, p. 285 [230].

[3] J. Clerk Maxwell, *A Dynamical Theory of the Electromagnetic Field* (SCIENTIFIC PAPERS, vol. I, p. 557 [458]).

[4] [p. 195].

[5] [vol. II, p. 231].

7.2 The Electrotonic State and the Magnetic Potential in the Memoir: ON FARADAY'S LINES OF FORCE

The group of three equations that we have studied does not alone constitute all of the electromagnetics of Maxwell. It is complemented by a series of key propositions. The form of these propositions, the series of deductions and inductions which [138] it provides them, vary from one writing to another; we must therefore analyze successively each of the memoirs composed on electricity by the Scottish physicist. In chronological order, we will start with the memoir entitled: *On Faraday's Lines of Force*.

In this memoir, as in his other writings prior to the *Treatise on Electricity and Magnetism*, Maxwell never takes into account the surfaces of discontinuity that the system can have; it is thus necessary, to follow his thoughts, to assume that two distinct media are always linked by a very thin but continuous *layer of passage*. It suffices that the remark was made so that any difficulty is excluded on this side.

It is not the same as the difficulties caused by clerical errors in calculation and, particularly, by the sign errors; they are constant in the passage which we propose to analyze and cast some uncertainty on the thinking of the author.

To the components α, β, γ of the *magnetic field* that he calls sometimes *magnetic force*, sometimes *magnetic intensity*, and sometimes *effective magnetizing force*, Maxwell adds another quantity, with components A, B, C,[6] that he calls *magnetic induction*; this word which, in more recent writings, will take another sense, means surely here the quantity usually considered in the theory of magnetism under the name of *intensity of magnetization*. In accordance with the ideas of Poisson, it should be [1st Part, equality (2.2)]

$$\frac{\partial A}{\partial x} + \frac{\partial B}{\partial y} + \frac{\partial C}{\partial z} = -\rho, \tag{7.6}$$

ρ being the density of the fictitious magnetic fluid, which Maxwell names *real magnetic matter*.[7] [139]

Between the magnitudes A, B, C and components α, β, γ of the field exist the relations

$$A = \frac{\alpha}{K}, \quad B = \frac{\beta}{K}, \quad C = \frac{\gamma}{K}, \tag{7.7}$$

where K is the resistance to magnetic induction[8]; if we continue to reconcile the theory of Maxwell's theory of Poisson, it is recognized that this resistance is the inverse of the coefficient of magnetization.

---

[6] We do not here keep notations of Maxwell.

[7] J. Clerk Maxwell, *On Faraday's Lines of Force* (SCIENTIFIC PAPERS, vol I, p. 192). Actually, instead of ρ, Maxwell wrote $4\pi\rho$; in addition, in the passage indicated, the sign of the second member of equality (7.6) is changed; but it is restored on p. 201.

[8] J. Clerk Maxwell, *loc. cit.*, p. 192.

Let V be a continuous function, zero to infinity, which is defined in the equation

$$\Delta V + 4\pi\rho = 0. \tag{7.8}$$

This function will be nothing other than the *magnetic potential function* that Poisson introduced into physics. Consider the differences

$$\begin{cases} a = A - \dfrac{1}{4\pi}\dfrac{\partial V}{\partial x}, \\[2mm] b = B - \dfrac{1}{4\pi}\dfrac{\partial V}{\partial y}, \\[2mm] c = C - \dfrac{1}{4\pi}\dfrac{\partial V}{\partial z}. \end{cases} \tag{7.9}$$

According to equalities (7.6) and (7.8), these differences will satisfy the relationship

$$\frac{\partial a}{\partial x} + \frac{\partial b}{\partial y} + \frac{\partial c}{\partial z} = 0. \tag{7.10}$$

Now, a theorem of analysis, often employed by Stokes, Helmholtz, and W. Thomson, shows that to three functions a, b, c, [140] related by relation (7.10), one can always associate three other functions F, G, H, such that we have

$$\begin{cases} a = -\dfrac{1}{4\pi}\left(\dfrac{\partial H}{\partial y} - \dfrac{\partial G}{\partial z}\right), \\[2mm] b = -\dfrac{1}{4\pi}\left(\dfrac{\partial F}{\partial z} - \dfrac{\partial H}{\partial x}\right), \\[2mm] c = -\dfrac{1}{4\pi}\left(\dfrac{\partial G}{\partial x} - \dfrac{\partial F}{\partial y}\right), \end{cases} \tag{7.11}$$

and

$$\frac{\partial F}{\partial x} + \frac{\partial G}{\partial y} + \frac{\partial H}{\partial z} = 0. \tag{7.12}$$

Therefore, equalities (7.9) can be written

$$\begin{cases} 4\pi A = \dfrac{\partial V}{\partial x} - \left(\dfrac{\partial H}{\partial y} - \dfrac{\partial G}{\partial z}\right), \\[2mm] 4\pi B = \dfrac{\partial V}{\partial y} - \left(\dfrac{\partial F}{\partial z} - \dfrac{\partial H}{\partial x}\right), \\[2mm] 4\pi C = \dfrac{\partial V}{\partial z} - \left(\dfrac{\partial G}{\partial x} - \dfrac{\partial F}{\partial y}\right). \end{cases} \tag{7.13}$$

Borrowing a name whereby Faraday referred to a rather vague concept, Maxwell[9] gives to the quantities F, G, H the name of components of the electrotonic state at the point (x, y, z).

What will the physical role attributed to these quantities be? The study of the electromagnetic potential of a system will tell us.

Let us return to Eq. (7.4).

In a system that contains no current, where, as a result, u, v, w are all equal to 0, these equations show us that the components α, β, γ of the magnetic field are the three partial derivatives of the same function; what is this [141] function? Guided by the classical theory, Maxwell admits[10] that it is the function $-V$, such that, in a system that contains magnets and current, we have

$$\alpha = -\frac{\partial V}{\partial x}, \quad \beta = -\frac{\partial V}{\partial x}, \quad \gamma = -\frac{\partial V}{\partial x}. \tag{7.14}$$

When a system of magnets moves, the forces exerted in this system, in accordance with the classical laws that Maxwell admits and that reflect previous equalities, perform some work; according to a well-known theorem, this work is the decrease incurred by the expression

$$\frac{1}{2} \int V \rho \, d\omega,$$

where the integral extends to all elements of volume $d\omega$ of the system. Why does Maxwell[11] omit the factor $1/2$ and write these lines: "...the whole work done during any displacement of a magnetic system is equal to the decrement of the integral

$$E = \int V \rho \, d\omega \tag{7.15}$$

throughout the system...[which we] now call...the *total potential of the system on itself*"? We see no reason. The fact remains that it would be impossible to correct this mistake and restore to E its true value without ruining, by the same token, any deduction that we want to analyze. So let us pass sentence on this error and continue.

Equality (7.15) can still be written, in virtue of equality (7.6),

$$E = -\int V \left(\frac{\partial A}{\partial x} + \frac{\partial B}{\partial y} + \frac{\partial C}{\partial z} \right) d\omega$$

or

$$E = \int \left(A \frac{\partial V}{\partial x} + B \frac{\partial V}{\partial y} + C \frac{\partial V}{\partial z} \right) d\omega$$

---

[9] J. Clerk Maxwell, *loc. cit.*, p. 203; the quantities F, G, H are designated by α_0, β_0, γ_0.

[10] J. Clerk Maxwell, *loc. cit.*, p. 202. In fact, in this passage, Maxwell said V; but on the next page, he restores a correct sign.

[11] J. Clerk Maxwell, *loc. cit.*, p. 203.

[142] or finally,[12] in virtue of equalities (7.14),

$$E = -\int (A\alpha + B\beta + C\gamma)d\omega. \tag{7.16}$$

Maxwell admits[13] that this expression of the potential extends to the case where the system contains not only magnets, but also currents. Then making use of equalities (7.13), equality (7.16) can be written

$$E = -\frac{1}{4\pi}\int \left(\alpha\frac{\partial V}{\partial x} + \beta\frac{\partial V}{\partial y} + \gamma\frac{\partial V}{\partial z}\right)d\omega$$

$$+\frac{1}{4\pi}\int \left[\left(\frac{\partial H}{\partial y} - \frac{\partial G}{\partial z}\right)\alpha + \left(\frac{\partial F}{\partial z} - \frac{\partial H}{\partial x}\right)\beta + \left(\frac{\partial G}{\partial x} - \frac{\partial F}{\partial y}\right)\gamma\right]d\omega. \tag{7.17}$$

We easily find, in virtue of equalities (7.14) and (7.8),

$$\int \left(\alpha\frac{\partial V}{\partial x} + \beta\frac{\partial V}{\partial y} + \gamma\frac{\partial V}{\partial z}\right)d\omega = -\int \left[\left(\frac{\partial V}{\partial x}\right)^2 + \left(\frac{\partial V}{\partial x}\right)^2 + \left(\frac{\partial V}{\partial x}\right)^2\right]d\omega$$

$$= \int V\Delta V d\omega = -4\pi\int V\rho d\omega.$$

Secondly, taking into account equalities (7.4), we find

$$\int \left[\left(\frac{\partial H}{\partial y} - \frac{\partial G}{\partial z}\right)\alpha + \left(\frac{\partial F}{\partial z} - \frac{\partial H}{\partial x}\right)\beta + \left(\frac{\partial G}{\partial x} - \frac{\partial F}{\partial y}\right)\gamma\right]d\omega$$

$$= \int \left[\left(\frac{\partial \gamma}{\partial y} - \frac{\partial \beta}{\partial z}\right)F + \left(\frac{\partial \alpha}{\partial z} - \frac{\partial \gamma}{\partial x}\right)G + \left(\frac{\partial \beta}{\partial x} - \frac{\partial \alpha}{\partial y}\right)H\right]d\omega$$

$$= -4\pi\int (Fu + Gv + Hw)d\omega.$$

[143] Equality (7.17) therefore becomes[14]

$$E = \int V\rho\,d\omega - \int (Fu + Gv + Hw)d\omega. \tag{7.18a}$$

Having reached this formula, Maxwell proposes to derive from the principle of conservation of energy the laws of electromagnetic induction, imitating, as he

---

[12]J. Clerk Maxwell, *loc. cit.*, p. 203, changes the sign of the second member.
[13]J. Clerk Maxwell, *loc. cit.*, p. 203.
[14]J. Clerk Maxwell, *loc. cit.*, p. 203.

acknowledges,[15] the well-known reasoning of Helmholtz in his memoir: *Ueber die Erhaltung der Kraft.*[16]

Imagine, he says, that external causes produce currents in the system. These causes *provide* work in two forms.

In the first place, they overcome the resistance that the conductors oppose to the passage of electricity; if we designate it by E_x, E_y, E_z, the components of the electromotive field at a point, the work provided for this purpose, during the time dt, is

$$-dt \int (E_x u + E_y v + E_z w) d\omega.$$

Secondly, they provide mechanical work that puts the system into motion; the work thus provided during the time dt is, by hypothesis, equal to the increase in the amount Q during the same time. Without justifying the omission of the term $\int V\rho \, d\omega$, Maxwell reduced[17] this increase to

$$-dt \frac{d}{dt} \int (Fu + Gv + Hw) d\omega$$

or again, assuming u, v, w invariant, to

$$-dt \int \left(\frac{dF}{dt} u + \frac{dG}{dt} v + \frac{dH}{dt} w \right) d\omega.$$

[144] If we suppose that the external causes disappear, and that currents are exclusively generated by the induction that the system exerts on itself, the work provided by these external causes must equal 0, where the equality

$$dt \int \left(\frac{dF}{dt} u + \frac{dG}{dt} v + \frac{dH}{dt} w \right) d\omega + dt \int (E_x u + E_y v + E_z w) d\omega = 0,$$

which can also be written

$$\int \left[\left(\frac{dF}{dt} + E_x \right) u + \left(\frac{dG}{dt} + E_y \right) v + \left(\frac{dH}{dt} + E_z \right) w \right] d\omega = 0. \qquad (7.19)$$

Equality (7.19) is satisfied if

$$E_x = -\frac{dF}{dt}, \quad E_y = -\frac{dG}{dt}, \quad E_z = -\frac{dH}{dt}. \qquad (7.20)$$

---

[15] J. Clerk Maxwell, *loc. cit.*, p. 204.

[16] [English translation: Brush and Hall (2003, 90–110)].

[17] J. Clerk Maxwell, *loc. cit.*, p. 204.

These equalities, of which Maxwell[18] admits the accuracy, connect the components of the electromotive induction field to the components of the electrotonic state.

7.3 Review of the Previous Theory

Do these equalities agree with the known laws of induction?

Maxwell has not given the analytical expression of the functions F, G, H any more than the function V and, therefore, has not developed equalities (7.20); but it is easy to make up for his silence.

The function V is, according to his oft-repeated feeling, the magnetic potential function, given by the equality

$$V(x, y, z) = \int \left(A_1 \frac{\partial \frac{1}{r}}{\partial x_1} + B_1 \frac{\partial \frac{1}{r}}{\partial y_1} + C_1 \frac{\partial \frac{1}{r}}{\partial z_1} \right) d\omega_1. \tag{7.21}$$

[145] Therefore, the conditions imposed on the functions F, G, H determine them unambiguously and give:

$$\begin{cases} F(x, y, z) = \int \left(C_1 \frac{\partial \frac{1}{r}}{\partial y_1} - B_1 \frac{\partial \frac{1}{r}}{\partial z_1} \right) d\omega_1, \\[3mm] G(x, y, z) = \int \left(A_1 \frac{\partial \frac{1}{r}}{\partial z_1} - C_1 \frac{\partial \frac{1}{r}}{\partial x_1} \right) d\omega_1, \\[3mm] H(x, y, z) = \int \left(B_1 \frac{\partial \frac{1}{r}}{\partial x_1} - A_1 \frac{\partial \frac{1}{r}}{\partial y_1} \right) d\omega_1. \end{cases} \tag{7.22}$$

If, in Eq. (7.20), one plugs in these expressions of the functions F, G, H, one finds, for the electromotive field components, expressions that very exactly agree with the known laws in the case where induction is produced by a change in magnetization, without which the system experiences no movement. The agreement is less perfect when the magnets and conductors move; a term is missing, that indeed it would easily be restored by ceasing to treat u, v, w as invariable and leaving constant only the electric flux of which these three quantities are components.

But a more serious objection stands against Maxwell's theory.

If this theory, applied to a moving system, denotes the existence of electromotive forces of induction, these electromotive forces all have this character of canceling themselves out when the system contains no magnet; the movement of conductors with currents passing through them would therefore be unable to cause any phenomenon of induction.

---

[18]J. Clerk Maxwell, *loc. cit.*, p. 204.

This single consequence suffices to condemn the theory described by Maxwell in his essay: *On Faraday's Lines of Force*.

We add a comment without which the reader would experience some trouble in comparing the previous formulas to those of Maxwell.

In the first place, Maxwell, in writing equalities (7.4), omits in the second member the factor 4π; he introduces this factor 4π instead in the second member of equality (7.6), and we need to indicate briefly how illogical this introduction is. [146]

Its starting point is what, in the memoir in question, Maxwell says regarding electric currents.[19]

If u, v, w are the components of the current at a point S on a closed surface, the amount of electricity that enters this surface during the time dt is

$$dt \int [u \cos (N_i, x) + v \cos (N_i, y) + w \cos (N_i, z)] dS.$$

Integration by parts transforms this expression into

$$-dt \int \left(\frac{\partial u}{\partial x} + \frac{\partial v}{\partial y} + \frac{\partial w}{\partial z} \right) d\omega, \tag{7.23a}$$

the integral extending over the volume that bounds the closed surface. By an evident sign error, Maxwell writes

$$+dt \int \left(\frac{\partial u}{\partial x} + \frac{\partial v}{\partial y} + \frac{\partial w}{\partial z} \right) d\omega. \tag{7.23b}$$

If e designates the electric density at a point inside surface S, the integral (7.23a) must be equal to

$$dt \int \frac{\partial e}{\partial t} d\omega,$$

which immediately gives the continuity equation

$$\frac{\partial u}{\partial x} + \frac{\partial v}{\partial y} + \frac{\partial w}{\partial z} + \frac{\partial e}{\partial t} = 0. \tag{7.24}$$

Maxwell does not write this equality, but he does write equality[20]

$$\frac{\partial u}{\partial x} + \frac{\partial v}{\partial y} + \frac{\partial w}{\partial z} = 4\pi\rho \tag{7.25}$$

without any explanation, if not that ρ is zero in the case of uniform currents. [147]

---

[19] J. Clerk Maxwell, *loc. cit.*, pp. 191–192.
[20] J. Clerk Maxwell, *loc. cit.*, p. 192, equality (C).

He is obviously free to consider a quantity ρ defined by this equality; this quantity ρ will be be equal to $-\frac{1}{4\pi}\frac{\partial e}{\partial t}$. Unfortunately, Maxwell seems to assume that the quantity ρ is precisely equal to $\frac{\partial e}{\partial t}$ and reasons accordingly; it is likely that this hypothesis guides him during the assimilation he establishes[21] between electrical conductivity and magnetization and leads him to connect the components of the magnetic induction to the magnetic density by the equality

$$\frac{\partial A}{\partial x} + \frac{\partial B}{\partial y} + \frac{\partial C}{\partial z} = 4\pi\rho, \tag{7.26}$$

which he replaces, some pages further,[22] by

$$\frac{\partial A}{\partial x} + \frac{\partial B}{\partial y} + \frac{\partial C}{\partial z} = -4\pi\rho. \tag{7.27}$$

We shall have occasion later to return to equality (7.26). For the moment, let us just note that the use of equalities (7.4) and (7.6) in the form we gave provides formulas which, sometimes, differ from those of Maxwell by the introduction or removal of a factor of 4π; but this amendment does not alter, we believe, the spirit of the theory.

It is, however, a last objection that could address the given interpretation of this theory. We have accepted without discussion that the *magnetic induction* of which Maxwell speaks should be identified here with the intensity of magnetization as it has been defined at the beginning of this work; that, therefore, the *magnetic resistance* K was the inverse of the *coefficient of magnetization k* considered by Poisson. This assimilation needs to be discussed.

On the surface that separates a magnet and a non-magnetic medium, [148] the magnetic potential function v satisfies the relation [1st Part, Chap. 2, equality (2.5)]

$$\frac{\partial V}{\partial N_i} + \frac{\partial V}{\partial N_e} = 4\pi[A\cos(N_i, x) + B\cos(N_i, y) + C\cos(N_i, z)],$$

N_i and N_e being the normal directions inwards and outwards from the magnet. If the laws of magnetism are those given by Poisson [*Ibid.*, equalities (2.6)], the second member of the previous equality becomes $-4\pi k\frac{\partial K}{\partial N_i}$, so that previous equality becomes

$$\frac{\partial K}{\partial N_e} + (1 + 4\pi k)\frac{\partial V}{\partial N_i} = 0. \tag{7.28}$$

[21]J. Clerk Maxwell, *loc. cit.*, p. 180.
[22]J. Clerk Maxwell, *loc. cit.*, p. 201.

Now, Maxwell clearly shows[23] that the magnetic resistance K is equal to the ratio

$$-\frac{\frac{\partial V}{\partial N_i}}{\frac{\partial V}{\partial N_e}}.$$

One must therefore put

$$K = \frac{1}{1 + 4\pi k}. \tag{7.29}$$

The quantity

$$\mu = 1 + 4\pi k \tag{7.30}$$

is what W. Thomson[24] called the *magnetic permeability*.

The electrical resistance Maxwell considers must therefore be taken equal to the inverse not of the coefficient of magnetization of Poisson, but of the magnetic permeability of W. Thomson.

The components of the *magnetic induction* are obtained by dividing [149] components α, β, γ of the field by the magnetic resistance or, what amounts to the same, by multiplying by the magnetic permeability. The expressions of these quantities are

$$A = (1 + 4\pi k)\alpha, \quad B = (1 + 4\pi k)\beta, \quad C = (1 + 4\pi k)\gamma, \tag{7.31}$$

while the components A, B, C of the *magnetization* have the values

$$\mathcal{A} = k\alpha, \quad \mathcal{B} = k\beta, \quad \mathcal{C} = k\gamma. \tag{7.32}$$

Magnetic induction and magnetization are not the same; their components are bound by the equalities

$$A = \frac{1 + 4\pi k}{k}\mathcal{A}, \quad B = \frac{1 + 4\pi k}{k}\mathcal{B}, \quad C = \frac{1 + 4\pi k}{k}\mathcal{C}. \tag{7.33}$$

When, therefore, we have identified the magnetic induction of Maxwell with the intensity of magnetization, we have committed a serious confusion.

If we committed it, it is because it seemed consistent with the thought of Maxwell, and that the developed theory seemed intimately connected to this confusion.

Certainly, in the memoir that we are analyzing, Maxwell did not perceive the distinction on which we insist; he proclaims[25] the complete mathematical identity of the formulas to which the classical theory of the magnetic polarity and the formulas supplied by his theory of the propagation by conduction of the lines of magnetic force led. Repeatedly, during his arguments, he carries the properties known about

---

[23] J. Clerk Maxwell, *loc. cit.*, p. 179.

[24] W. Thomson, PAPERS ON ELECTROSTATICS AND MAGNETISM, art. 629; 1872.

[25] J. Clerk Maxwell, *On Faraday's Lines of Force* (SCIENTIFIC PAPERS, vol. I, p. 179).

magnetization over to *magnetic induction*. In particular, the point here seems to be very clear:

The confusion between the concept of *magnetic induction* and the concept of intensity of magnetization that is considered in the classical theory of magnetism has only led Maxwell when he determined a relationship between the variations that the magnetic induction [150] feels from one point to another and the density of the *magnetic material*. When, in his *Treatise on Electricity and Magnetism*, Maxwell will manage to distinguish the two concepts of intensity of magnetization and magnetic induction, he will not establish any relationship between the derivatives of the components of the latter and the magnetic density.

7.4 The Electrotonic State and the Electromagnetic Energy in the Memoir: ON PHYSICAL LINES OF FORCE

Our intention is not to discuss here the mechanical problems that the theory outlined in the memoir: *On Physical Lines of Force* poses. Accepting as demonstrated all the dynamical laws that Maxwell states regarding the medium that he has imagined, we will examine only how Maxwell carries these laws from the field of mechanics to the field of electricity.

The fluid contained in the cells is driven by a whirling motion; let, at the point (x, y, z) and time t, α, β, γ be the projections on the axes of a segment equal to the angular speed of rotation and focused on the instantaneous axis of rotation of the element $d\omega$; let, in addition, μ be a quantity proportional to the density of the fluid that drives these vortical movements. According to Maxwell, an element of volume $d\omega$ of fluid is subjected to a force of which $X\,d\omega$, $Y\,d\omega$, $Z\,d\omega$ are the components. X has the following form[26]:

$$
\begin{aligned}
X = &\frac{1}{4\pi}\left(\frac{\partial}{\partial x}\mu\alpha + \frac{\partial}{\partial y}\mu\beta + \frac{\partial}{\partial z}\mu\gamma\right)\alpha + \frac{\mu}{8\pi}\frac{\partial(\alpha^2 + \beta^2 + \gamma^2)}{\partial x} \\
&+ \frac{\mu\gamma}{4\pi}\left(\frac{\partial\gamma}{\partial x} - \frac{\partial\alpha}{\partial z}\right) - \frac{\mu\beta}{4\pi}\left(\frac{\partial\alpha}{\partial y} - \frac{\partial\beta}{\partial x}\right) - \frac{\partial\Pi}{\partial x}.
\end{aligned}
\tag{7.34}
$$

Y et Z have analogous expressions.

Leaving aside the term $-\frac{\Pi}{x}$, where Π represents a certain pressure, Maxwell strives to give an electromagnetic interpretation [151] of the other terms that form the second member of equality (7.34).

The starting point of this interpretation is the following:

The magnitudes α, β, γ, components of rotation, represent the components of the magnetic field at each point.

---

[26]J. Clerk Maxwell, *On Physical Lines of Force*, SCIENTIFIC PAPERS, vol. I, p. 458.

Therefore, if the element $d\omega$ contains a mass m of magnetic fluid, it must be subjected to a force with components αm, βm, γm. Among the terms that form X, we must, in the first place, according to Maxwell, find the term $\alpha\rho$, $\rho = \frac{m}{d\omega}$ being the density of the magnetic fluid at the point under consideration, and this term can only be the first. Maxwell is also led to admit that the density of magnetic fluid at a point is given by the equality

$$\frac{\partial}{\partial x}\mu\alpha + \frac{\partial}{\partial y}\mu\beta + \frac{\partial}{\partial z}\mu\gamma = 4\pi\rho. \tag{7.35a}$$

Maxwell, who again called the quantities $\mu\alpha$, $\mu\beta$, $\mu\gamma$ the *components of the magnetic induction*, is thus led to take up again equality (7.26) which he had proposed, then abandoned, in his previous memoir.

Does Maxwell seek to justify this relationship otherwise than by needing to find a certain term for the second member of equality (7.34)? He only writes in this sense these few lines[27]:

...so...

$$\left(\frac{\partial}{\partial x}\mu\alpha + \frac{\partial}{\partial y}\mu\beta + \frac{\partial}{\partial z}\mu\gamma\right)d\omega = 4\pi\rho\, d\omega,$$

which represents the total amount of magnetic induction outwards through the surface of the element $d\omega$,[28] represents the amount of "imaginary magnetic matter" within the element, of the kind which points north.

But these lines are contrary to the purpose pursued by Maxwell, because they would lead to writing $\rho\, d\omega$ for the second member, and not $4\pi\rho d\omega$. [152]

The influence on the mind of Maxwell by the strange equality (7.25), written in his previous memoir, is very tangible here.

If α, β, γ represent the components of the magnetic field, the components u, v, w of the electric current must satisfy equalities (7.4). For the second member of X, we have

$$\frac{\mu}{4\pi}\left(\frac{\partial\gamma}{\partial x} - \frac{\partial\alpha}{\partial z}\right) - \frac{\mu\beta}{4\pi}\left(\frac{\partial a}{\partial y} - \frac{\partial\beta}{\partial x}\right) = \mu(\gamma v - \beta w), \tag{7.36}$$

which would represent the parallel component to Ox of the electromagnetic action.

It remains to interpret the term

$$\frac{\mu}{8\pi}\frac{\partial}{\partial x}(\alpha^2 + \beta^2 + \gamma^2). \tag{7.37}$$

It represents the component parallel to Ox of a force that tends to lead the element $d\omega$ to the region of space where the field has the largest absolute value. Faraday[29] had already shown that you could regard a small diamagnetic body, i.e. a body for which

---

[27] J. Clerk Maxwell, *loc. cit.*, p. 459.

[28] [$dx\, dy\, dz$ in the original of Maxwell].

[29] Faraday, *Experimental Researches*, §2418 PHILOSOPHICAL TRANSACTIONS, 1846, p. 21.

μ is lower than in the surrounding medium, as if it were directed toward the region of space where the field has lesser absolute value; and W. Thomson had shown[30] that a small, perfectly soft body was somehow attracted to the point of space where $(\alpha^2 + \beta^2 + \gamma^2)$ has the largest value. Maxwell does not hesitate to show in the term (7.37) the component of this attraction parallel to Ox.

But a serious objection can be made to this interpretation.

When a perfectly soft body is subjected to magnetic induction, the magnetization it takes can be represented by a certain distribution of magnetic fluid; the actions it undergoes can be decomposed into forces that would act on the various elementary masses of magnetic fluid. The apparent attraction exerted on the perfectly soft body by the point where the [153] field reaches its largest absolute value is not an action separate from the preceding and superposed on the previous ones; it is only the result. The interpretation of Maxwell makes him find twice, for the second member of equality (7.34), an action that the recognized laws of magnetism admit only once.

This difficulty is not the only one that faces the theory which we continue to present.

Suppose[31] that the system contains no electric current; the equalities, then verified,

$$u = 0, \quad v = 0, \quad w = 0,$$

will be transformed, according to equalities (7.4), into

$$\frac{\partial \gamma}{\partial y} - \frac{\partial \beta}{\partial z} = 0, \quad \frac{\partial \alpha}{\partial z} - \frac{\partial \gamma}{\partial x} = 0, \quad \frac{\partial \beta}{\partial x} - \frac{\partial \alpha}{\partial y} = 0;$$

the components α, β, γ of the magnetic field will be the three partial derivatives of the same function:

$$\alpha = -\frac{\partial V}{\partial x}, \quad \beta = -\frac{\partial V}{\partial y}, \quad \gamma = -\frac{\partial V}{\partial z}. \tag{7.38}$$

Equality (7.35a) will become

$$\frac{\partial}{\partial x}\left(\mu \frac{\partial V}{\partial x}\right) + \frac{\partial}{\partial y}\left(\mu \frac{\partial V}{\partial y}\right) + \frac{\partial}{\partial z}\left(\mu \frac{\partial V}{\partial z}\right) = -4\pi\rho, \tag{7.39}$$

and, in a region where μ does not change value when one moves from one point to the next point,

$$\Delta V = -4\pi \frac{\rho}{\mu}. \tag{7.40}$$

---

[30] W. Thomson, Philosophical Magazine, October 1850.—Papers on Electrostatics and Magnetism, n° 647.

[31] J. Clerk Maxwell, loc. cit., p. 464.

Imagine that μ has the same value throughout all of space. Suppose that a region 1 of this space contains the "imaginary [154] magnetic material" such that ρ differs from 0, while ρ is zero in all the rest of the space; we will have

$$V = \frac{1}{\mu} \int_1 \frac{\rho_1}{r} d\omega_1. \tag{7.41}$$

The magnetic field will therefore be calculated as if two masses m, m', located at the distance of r, are repelled with a force

$$\frac{1}{\mu} \frac{mm'}{r^2}.$$

In the vacuum where, by definition, $\mu = 1$, this force has the expression $\frac{mm'}{r^2}$ given by Coulomb, of which it also seems we have found the law, a conclusion however that we should not rush to affirm, because the previous deduction is subject to the hypothesis that μ has the same value within the magnetized masses and the interposed medium, an unacceptable hypothesis when it comes to iron masses placed in air.

We add this remark, well able to discredit any theory of magnetism given by Maxwell. According to the classical theory, any magnet still contains as much boreal magnetic fluid as austral magnetic fluid; so the total magnetic charge it contains is always equal to 0. This conclusion no longer has force in the theory of Maxwell; such that, according to this theory, it seems possible to isolate a magnet that would contain only the boreal fluid or only the austral fluid.

The previous considerations play a large role in the determination of the form that should be attributed to magnetic energy.[32]

The fluid, animated with vortical movements representing the magnetic field, has a certain live force[33]; this live force has the value

$$E = C \int \mu(\alpha^2 + \beta^2 + \gamma^2) d\omega, \tag{7.42}$$

[155] the integral extending over the entire system and C being a constant coefficient whose value is to be determined.

To achieve this, Maxwell assumes that the system contains no current, in which case equalities (7.38) are applicable. Equality (7.42) then becomes

$$E = C \int \mu \left[\left(\frac{\partial V}{\partial x}\right)^2 + \left(\frac{\partial V}{\partial y}\right)^2 + \left(\frac{\partial V}{\partial z}\right)^2 \right] d\omega.$$

---

[32]J. Clerk Maxwell, *loc. cit.*, p. 472.
[33][*Force vive* or mv^2, related to the kinetic energy $mv^2/2$].

He next assumes that μ has the same value in all space, which allows him to transform the previous equality into

$$E = -C \int \mu V \Delta V \, d\omega.$$ (7.43)

He finally assumes that the function V is the sum of two functions:

$$V = V_1 + V_2.$$

The first, V_1, satisfies, at any point in the volume ω_1, the equality

$$\Delta V_2 = -\frac{4\pi \rho_2}{\mu},$$

and, at any other point, equality $\Delta V_1 = 0$. The second, V_2, satisfies, at any point in a volume ω_2 not having any point in common with ω_1, the equality

$$\Delta V_2 = -\frac{4\pi \rho_2}{\mu}$$

and, at any other point, the equality $\Delta V_2 = 0$. Therefore, equality (7.43) can be written

$$E = 4\pi C \int_{\omega_1} (V_1 + V_2) \rho_1 \, d\omega_1 + 4\pi C \int_{\omega_2} (V + V_2) \rho_2 \, d\omega_2.$$ (7.44)

Furthermore, Green's theorem gives the equality

$$\int V_1 \Delta V_2 \, d\omega = \int V_2 \Delta V_1 \, d\omega,$$

[156] where the integrals extend over the whole space. This equality is easily transformed into the following,

$$\int_{\omega_2} V_1 \rho_2 \, d\omega_2 = \int_{\omega_1} V_2 \rho_1 \, d\omega_1,$$

which transforms equality (7.44) into

$$E = 4\pi C \int_{\omega_1} V_1 \rho_1 \, d\omega_1 + 4\pi C \int_{\omega_2} V_2 \rho_2 \, d\omega_2 + 8\pi C \int_{\omega_2} V_1 \rho_2 \, d\omega_2.$$ (7.45)

Suppose the volume ω_1 and the value of ρ_1, which corresponds to each of its points, remain fixed. Suppose the volume ω_2 moves as a rigid solid, each of its points leading to the value of ρ_2, which corresponds to it. We easily recognize that $\int_{\omega_1} V_1 \rho_1 \, d\omega_1$ and $\int V_2 \rho_2 \, d\omega_2$ will keep the values invariable, while if we mean by

∂x_2, ∂y_2, ∂z_2 the components of the displacement of a point of the element $d\omega_2$, we will have

$$\partial \int_{\omega_2} V_1 \rho_2 \, d\omega_2 = \int_{\omega_2} \rho_2 \left(\frac{\partial V_1}{\partial x_2} \partial x_2 + \frac{\partial V_1}{\partial y_2} \partial y_2 + \frac{\partial V_1}{\partial z_2} \partial z_2 \right) d\omega_2$$

and

$$\partial E = 8\pi C \int_{\omega_2} \rho_2 \left(\frac{\partial V_1}{\partial x_2} \partial x_2 + \frac{\partial V_1}{\partial y_2} \partial y_2 + \frac{\partial V_1}{\partial z_2} \partial z_2 \right) d\omega_2. \tag{7.46}$$

This variation of the energy must be equal and opposite in sign to the work of *apparent* forces that the magnet ω_1 exerts on the magnet ω_2.

Taking account of the first term of expression (7.34) of X and of the interpretation that he gave, but completely forgetting the second term, Maxwell admits that this work has the value

$$dT = \int_{\omega_2} \rho_2 (\alpha_1 \, \partial x_2 + \beta_1 \, \partial y_2 + \gamma_1 \, \partial x_2) d\omega_2$$

[157] or else, by virtue of equalities (7.38),

$$dT = - \int_{\omega_2} \rho_2 \left(\frac{\partial V_1}{\partial x_2} \partial x_2 + \frac{\partial V_1}{\partial y_2} \partial y_2 + \frac{\partial V_1}{\partial z_2} \partial z_2 \right) d\omega_2.$$

By identifying the expression $-dT$ with the expression ∂E given by equality (7.46), we find

$$8\pi C = 1,$$

such that equality (7.42) becomes

$$E = \frac{1}{8\pi} \int \mu(\alpha^2 + \beta^2 + \gamma^2) d\omega. \tag{7.47a}$$

Thus, the expression of the live force or electromagnetic kinetic energy is obtained. This expression will play a significant role in the work of Maxwell.

Here is an important application.[34]

Imagine a stationary system where α, β, γ vary from one moment to the next. The system will be traversed by electrical currents generated by induction. The production of these currents corresponds to a certain increase of energy of the system; and Maxwell admits that if E_x, E_y, E_z are the components of the electromotive field,

---

[34]J. Clerk Maxwell, SCIENTIFIC PAPERS, vol. I, p. 475.

the increase of energy, within the system, in time dt, corresponding to the creation of the electrical currents, has the value

$$dt \int (E_x u + E_y v + E_z w) d\omega.$$

The total energy of the system, which is assumed to subtract from all external action, must remain invariant, the increase of which we just gave the expression should be offset by an equal decrease in the electromagnetic live force. This decrease has, moreover, the value

$$-\frac{dt}{4\pi} \int \mu \left(\alpha \frac{\partial \alpha}{\partial t} + \beta \frac{\partial \beta}{\partial t} + \gamma \frac{\partial \gamma}{\partial t} \right) d\omega.$$

[158] We will therefore have the equality

$$(E_x u + E_y v + E_z w) d\omega + \frac{1}{4\pi} \int \mu \left(\alpha \frac{\partial \alpha}{\partial t} + \beta \frac{\partial \beta}{\partial t} + \gamma \frac{\partial \gamma}{\partial t} \right) d\omega = 0. \qquad (7.48)$$

But, in virtue of equalities (7.4),

$$\int (E_x u + E_y v + E_z w) d\omega$$

$$= -\frac{1}{4\pi} \int \left[\left(\frac{\partial \gamma}{\partial y} - \frac{\partial \beta}{\partial z} \right) E_x + \left(\frac{\partial \alpha}{\partial z} - \frac{\partial \gamma}{\partial x} \right) E_y + \left(\frac{\partial \beta}{\partial x} - \frac{\partial \alpha}{\partial y} \right) E_z \right] d\omega$$

$$= -\frac{1}{4\pi} \int \left[\left(\frac{\partial E_z}{\partial y} - \frac{\partial E_y}{\partial z} \right) \alpha + \left(\frac{\partial E_x}{\partial z} - \frac{\partial E_z}{\partial x} \right) \beta + \left(\frac{\partial E_y}{\partial x} - \frac{\partial E_x}{\partial y} \right) \gamma \right] d\omega.$$

Equality (7.48) then becomes

$$\int \left[\left(\frac{\partial E_z}{\partial y} - \frac{\partial E_y}{\partial z} - \mu \frac{\partial \alpha}{\partial t} \right) \alpha + \left(\frac{\partial E_x}{\partial z} - \frac{\partial E_x}{\partial x} - \mu \frac{\partial \beta}{\partial t} \right) \beta \right.$$
$$\left. + \left(\frac{\partial E_y}{\partial x} - \frac{\partial E_x}{\partial y} - \mu \frac{\partial \gamma}{\partial t} \right) \gamma \right] d\omega = 0.$$

It will be obviously verified if, at each point,

$$\begin{cases} \dfrac{\partial E_z}{\partial y} - \dfrac{\partial E_y}{\partial z} = \mu \dfrac{\partial \alpha}{\partial t}, \\[2mm] \dfrac{\partial E_x}{\partial z} - \dfrac{\partial E_z}{\partial x} = \mu \dfrac{\partial \beta}{\partial t}, \\[2mm] \dfrac{\partial E_y}{\partial x} - \dfrac{\partial E_x}{\partial y} = \mu \dfrac{\partial \gamma}{\partial t}. \end{cases} \qquad (7.49)$$

The three equations that we just wrote are of great importance; together with the three Eq. (7.4), they form what it is convenient to name—with Heaviside, Hertz, and Cohn—the *six equations of Maxwell*.

Let $\Psi(x, y, z, t)$ be the function, defined for a function near t, which satisfies in all of space the relationship

$$\Delta\Psi + \frac{\partial E_x}{\partial x} + \frac{\partial E_y}{\partial y} + \frac{\partial E_z}{\partial z} = 0. \tag{7.50}$$

[159] We put

$$\begin{cases} E_x = -\dfrac{\partial\Psi}{\partial x} + E'_x, \\[2mm] E_y = -\dfrac{\partial\Psi}{\partial y} + E'_y, \\[2mm] E_z = -\dfrac{\partial\Psi}{\partial z} + E'_z. \end{cases} \tag{7.51}$$

Equalities (7.49) and (7.50) will become

$$\begin{cases} \dfrac{\partial E'_z}{\partial y} - \dfrac{\partial E'_y}{\partial z} = \mu\dfrac{\partial\alpha}{\partial t}, \\[2mm] \dfrac{\partial E'_x}{\partial z} - \dfrac{\partial E'_z}{\partial x} = \mu\dfrac{\partial\beta}{\partial t}, \\[2mm] \dfrac{\partial E'_y}{\partial x} - \dfrac{\partial E'_x}{\partial y} = \mu\dfrac{\partial\gamma}{\partial t}. \end{cases} \tag{7.52}$$

$$\frac{\partial E'_x}{\partial x} + \frac{\partial E'_y}{\partial y} + \frac{\partial E'_z}{\partial z} = 0. \tag{7.53}$$

These equations, verified throughout all of space, are treated by Maxwell in the following manner[35]:

Let F, G, H be three functions that satisfy in all of space relations

$$\begin{cases} \dfrac{\partial H}{\partial y} - \dfrac{\partial G}{\partial z} = -\mu\alpha, \\[2mm] \dfrac{\partial F}{\partial z} - \dfrac{\partial H}{\partial x} = -\mu\beta, \\[2mm] \dfrac{\partial G}{\partial x} - \dfrac{\partial F}{\partial y} = -\mu\gamma, \end{cases} \tag{7.54a}$$

---

[35]Indeed, in the analyzed passage, Maxwell designates by $-F$, $-G$, $-H$ the quantities which we refer to here as F, G, H; the change of sign we introduced restores the concordance among the various writings of Maxwell.

$$\frac{\partial F}{\partial x} + \frac{\partial G}{\partial y} + \frac{\partial H}{\partial z} = 0. \tag{7.55a}$$

[160] We will have

$$E_x = -\frac{\partial F}{\partial t}, \quad E_y = -\frac{\partial G}{\partial t}, \quad E_z = -\frac{\partial H}{\partial t},$$

and equalities (7.51) will become

$$\begin{cases} E_x = -\dfrac{\partial \Psi}{\partial x} - \dfrac{\partial F}{\partial t}, \\[2mm] E_y = -\dfrac{\partial \Psi}{\partial y} - \dfrac{\partial G}{\partial t}, \\[2mm] E_z = -\dfrac{\partial \Psi}{\partial z} - \dfrac{\partial H}{\partial t}. \end{cases} \tag{7.56a}$$

The functions F, G, H, which are contained in these formulas, are the components of the *electrotonic state*, already considered by Maxwell in his memoir: *On Faraday's Lines of Force*. As for Ψ, it[36]

> is a function of x, y, z, and t, which is indeterminate as far as regards the solution of the original equations, but which may always be determined in any given case from the circumstances of the problem. The physical interpretation of Ψ is, that it is the *electric tension* at each point of space.[37]

In a system where the steady state is established, F, G, H no longer depend on time; equalities (7.56a) reduce to

$$E_x = -\frac{\partial \Psi}{\partial x}, \quad E_y = -\frac{\partial \Psi}{\partial y}, \quad E_z = -\frac{\partial \Psi}{\partial z}. \tag{7.57}$$

The components of the electromotive field are respectively equal to three partial derivatives of a function whose analytical form remains absolutely unknown. It is one of the foundations of Maxwell's second electrostatics.[38]

By exposing this calculation, Maxwell notes very precisely[39] [161] that equations (7.54a) cannot be written if we have at any point

$$\frac{\partial}{\partial x}\mu\alpha + \frac{\partial}{\partial y}\mu\beta + \frac{\partial}{\partial z}\mu\gamma = 0. \tag{7.58a}$$

---

[36] In the study on induction in a stationary system, Maxwell has forgotten the terms in $-\frac{\partial \Psi}{\partial x}$, $-\frac{\partial \Psi}{\partial y}$, $-\frac{\partial \Psi}{\partial z}$; but he recovered them in the formulas pertaining to induction within a moving system.

[37] [t. I, p. 482].

[38] See 1st Part, Chap. 4.

[39] J. Clerk Maxwell, Scientific Papers, vol. I, p. 476, equality (57).

Brought closer to equality (7.35a), this latter equality becomes

$$\rho = 0.$$

Equation (7.54a) can be written only if the *fictitious magnetic material everywhere has zero density*. The theory of the electrotonic state that Maxwell developed here is incompatible with the existence of magnetism; it is a restriction that Maxwell will forget in his memoir: *A Dynamical Theory of the Electromagnetic Field*.

7.5 The Electrotonic State and Electromagnetic Energy in the Memoir: A DYNAMICAL THEORY OF THE ELECTROMAGNETIC FIELD

In the document entitled: *On Physical Lines of Force*, Maxwell has endeavoured to create a mechanical assemblage whose properties could be regarded as the explanation of electrical phenomena. In his later writings, while continuing to admit that electric and magnetic actions are essentially mechanical, he no longer seeks to build the machinery which produces them. According to the council of Pascal, he continues to "say in general: this happens through shape and motion;" but he no longer tries "to say which and compose the machine."[40] To formulate the expression of electrostatic energy and electromagnetic energy; to show that to these expressions one can attach the laws of electrical phenomena, imitating the Lagrange's method of deriving the equations of motion of a system from expressions of the potential and kinetic energies of the system; these are the objects of the memoir: *A Dynamical Theory of the Electromagnetic Field* and the *Treatise on Electricity and Magnetism*.

The third part of the memoir: *A Dynamical Theory of the Electromagnetic Field*, which interests us here, offers, [162] in an extremely concise form, the union of the main formulas governing electrical phenomena.

One of the quantities that Maxwell introduces, firstly, is the *electromagnetic moment*[41]; this vector, whose components he designates by F, G, H, plays exactly the role he attributed to it in his previous memoirs on the *electrotonic state*; he readily admits, in fact, that the components E'_x, E'_y, E'_z of the electromotive induction field in a stationary system are given by equalities

$$E'_x = -\frac{\partial F}{\partial t}, \quad E'_y = -\frac{\partial G}{\partial t} \quad E'_z = -\frac{\partial H}{\partial t}.$$

Maxwell gives no analytic expression to these quantities F, G, H, but he connects them to the components α, β, γ of the magnetic field. Designating the components

---

[40][Pascal (2004, p. 25, S118/L84)].

[41]J. Clerk Maxwell, SCIENTIFIC PAPERS, vol. I, p. 555.

of the magnetic induction by $\mu\alpha$, $\mu\beta$, $\mu\gamma$, he wrote the three relations[42]

$$\begin{cases} \dfrac{\partial H}{\partial y} - \dfrac{\partial G}{\partial z} = -\mu\alpha, \\[2mm] \dfrac{\partial F}{\partial z} - \dfrac{\partial H}{\partial x} = -\mu\beta, \\[2mm] \dfrac{\partial G}{\partial x} - \dfrac{\partial F}{\partial y} = -\mu\gamma. \end{cases} \tag{7.54a}$$

These equalities, verified throughout all of space, are exactly the same as equalities (7.54a); but to equalities (7.54a) the following relation is joined:

$$\frac{\partial F}{\partial x} + \frac{\partial G}{\partial y} + \frac{\partial H}{\partial z} = 0, \tag{7.55a}$$

such that the functions F, G, H were determined. In the memoir that we are currently analyzing, Maxwell no longer admits [163] the accuracy of equality (7.55a); on the contrary, he writes[43]

$$\frac{\partial F}{\partial x} + \frac{\partial G}{\partial y} + \frac{\partial H}{\partial z} = J, \tag{7.55b}$$

and he treats the quantity J as an unknown quantity, generally different from 0.

Therefore, the quantities F, G, H are no longer determined; you can add to them, respectively, the three derivatives with respect to x, y, z of an arbitrary function of variables x, y, z, t.

When, therefore, Maxwell wrote[44] the components of the electromotive field within a stationary system

$$\begin{cases} E_x = -\dfrac{\partial \Psi}{\partial x} - \dfrac{\partial F}{\partial t}, \\[2mm] E_y = -\dfrac{\partial \Psi}{\partial y} - \dfrac{\partial G}{\partial t}, \\[2mm] E_z = -\dfrac{\partial \Psi}{\partial z} - \dfrac{\partial H}{\partial t}, \end{cases} \tag{7.56b}$$

he can, in all circumstances, substitute for Ψ any function of x, y, z, t. The function Ψ is absolutely indeterminate and could logically agree with the following statement[45]:

> Ψ is a function of x, y, z, and t, which is indeterminate as far as regards the solution of the above equations, because the terms depending on it will disappear on integrating round the circuit. The quantity Ψ can always, however, be determined in any particular case when we know the actual conditions of the question. The physical interpretation of Ψ is, that it represents the *electric potential* at each point of space.

---

[42] J. Clerk Maxwell, *loc. cit.*, p. 556.

[43] J. Clerk Maxwell, SCIENTIFIC PAPERS, vol. I, p. 578.

[44] J. Clerk Maxwell, *loc. cit.*, p. 558 and p. 578.

[45] J. Clerk Maxwell, *loc. cit.*, p. 558.

In addition, when Maxwell, in his memoir: *On Physical Lines of Force*, had written Eq. (7.54a), he had taken care to note [164] that they would be absurd if on did not have, in all of space, the equality

$$\frac{\partial}{\partial x}\mu\alpha + \frac{\partial}{\partial y}\mu\beta + \frac{\partial}{\partial z}\mu\gamma = 0. \tag{7.58a}$$

In the present memoir, he fails to make this remark and, what is more, he reasons as if equality (7.58a) were wrong; we will see in time an example.

As a result of considerations[46] whose extreme brevity makes it difficult to consider as reasoning, Maxwell admits[47] that the electromagnetic energy is given by the formula

$$E = \frac{1}{2}\int (Fu + Gv + Hw)d\omega, \tag{7.59}$$

where u, v, w represent the components of the total current and where the integral extends over all of space.

We will seek to clarify the considerations which led Maxwell to this expression.

In time dt, the system releases, according to Joule's law, a quantity of heat given in *mechanical units* by the expression

$$dt \int r(u^2 + v^2 + w^2)d\omega,$$

where r is the specific resistance of the medium; in virtue of Ohm's law, this amount of heat can also be written

$$dt \int (E_x u + E_y v + E_z w)d\omega.$$

If the system, isolated and immobile, is subject only to the electromotive actions that the fluctuations in the flow of electricity produce by induction, this amount of heat output in [165] time dt is exactly equal to the reduction of electromagnetic energy during the same time; so we have the equality

$$dE + dt \int (E_x u + E_y v + E_z w)d\omega = 0.$$

At the same time, the components E_x, E_y, E_z of the electromotive field are given by equalities (7.56b), so that the previous equality becomes

---

[46] J. Clerk Maxwell, SCIENTIFIC PAPERS, vol. I, p. 541.
[47] J. Clerk Maxwell, *loc. cit.*, p. 562.

$$dE - dt \int \left(\frac{\partial \Psi}{\partial x} u + \frac{\partial \Psi}{\partial y} v + \frac{\partial \Psi}{\partial z} w \right) d\omega$$

$$- dt \int \left(\frac{\partial F}{\partial x} u + \frac{\partial G}{\partial y} v + \frac{\partial H}{\partial z} w \right) d\omega = 0.$$

The term

$$-dt \int \left(\frac{\partial \Psi}{\partial x} u + \frac{\partial \Psi}{\partial y} v + \frac{\partial \Psi}{\partial z} w \right) d\omega$$

can be written

$$\int \Psi \left(\frac{\partial u}{\partial x} + \frac{\partial v}{\partial y} + \frac{\partial w}{\partial z} \right) d\omega.$$

It is therefore equal to 0 if one considers only uniform currents for which

$$\frac{\partial u}{\partial x} + \frac{\partial v}{\partial y} + \frac{\partial w}{\partial z} = 0.$$

So we have

$$dE = dt \int \left(\frac{\partial F}{\partial t} u + \frac{\partial G}{\partial t} v + \frac{\partial H}{\partial t} w \right) d\omega. \tag{7.60}$$

Is this equality compatible with the expression of E that equality (7.59) provides? It gives equality

$$dE = \frac{1}{2} dt \int \left(\frac{\partial F}{\partial x} u + \frac{\partial G}{\partial y} v + \frac{\partial H}{\partial z} w \right)$$

$$+ \frac{1}{2} \left(F \frac{\partial u}{\partial x} + G \frac{\partial v}{\partial y} + H \frac{\partial w}{\partial z} \right) d\omega.$$

[166] So that this equality is compatible with equality (7.60), it is necessary and sufficient to have the equality

$$\int \left(\frac{\partial F}{\partial t} u + \frac{\partial G}{\partial t} v + \frac{\partial H}{\partial t} w \right) d\omega$$

$$= \int \left(F \frac{\partial u}{\partial t} + G \frac{\partial v}{\partial t} + H \frac{\partial w}{\partial t} \right) d\omega. \tag{7.61}$$

Is this equality satisfied? It is impossible to decide since, in the memoir that we are analyzing, Maxwell gives no determinate analytic expressions to the functions F, G, H.

We accept equality (7.59). Equalities (7.4) will give it the form

$$E = -\frac{1}{8\pi} \int \left[\left(\frac{\partial \gamma}{\partial y} - \frac{\partial \beta}{\partial z} \right) F + \left(\frac{\partial \alpha}{\partial z} - \frac{\partial \gamma}{\partial x} \right) G + \left(\frac{\partial \beta}{\partial x} - \frac{\partial \alpha}{\partial y} \right) H \right] d\omega$$

that an integration by parts will change into

$$E = -\frac{1}{8\pi} \int \left[\left(\frac{\partial H}{\partial y} - \frac{\partial G}{\partial z} \right) \alpha + \left(\frac{\partial F}{\partial z} - \frac{\partial H}{\partial x} \right) \beta + \left(\frac{\partial G}{\partial x} - \frac{\partial F}{\partial y} \right) \gamma \right] d\omega.$$

Equalities (7.54b) will then give

$$E = \frac{1}{8\pi} \int \mu(\alpha^2 + \beta^2 + \gamma^2) d\omega. \qquad (7.47a)$$

The electromagnetic energy, determined in the memoir: *A Dynamical Theory of the Electromagnetic Field* by electrical considerations, thus takes the form, in the memoir: *On Physical Lines of Force*, of mechanical hypotheses that he had attributed to it.

The agreement between these forms of electromagnetic energy and those to which Maxwell was led, in his memoir: *On Faraday's Lines of Force*, is more difficult to establish from the theory of magnetism.

The latter form is given by the equality

$$E = \int V\rho \, d\omega - \int (Fu + Gv + Hw) d\omega. \qquad (7.18b)$$

[167] The magnetic density ρ is related to the components of the magnetic induction by equality (7.35a)

$$\frac{\partial}{\partial x} \mu\alpha + \frac{\partial}{\partial y} \mu\beta + \frac{\partial}{\partial z} \mu\gamma = 4\pi\rho$$

or, in virtue of equalities (7.54b),

$$\rho = 0.$$

Equality (7.18a) therefore reduces to

$$E = -\int (Fu + Gv + Hw) d\omega. \qquad (7.62)$$

This expression of electromagnetic energy differs from expression (7.59) of the same quantity at once by the presence of the "−" sign and the lack of the factor $\frac{1}{2}$. In truth, as we have remarked in Sect. 7.2, the absence of the factor $\frac{1}{2}$ comes from an omission, and this factor could be easily restored. But the contradiction that the nature of the signs introduced between the two expressions of the electromagnetic energy cannot be avoided.

It would disappear, however, if in the definition of the magnetic density, given by equality (7.6), we changed the sign of ρ; Maxwell made this change in sign accidentally in the memoir: *On Faraday's Lines of Force*, then normally in his subsequent memoirs.

Does the expression (7.47a) of the electromagnetic energy agree with the laws known to magnetism? Whether the system does or does not contain currents, Maxwell admits that there exists a function Φ[48] which he called *magnetic potential*, such that one has

$$\alpha = -\frac{\Phi}{x}, \quad \beta = -\frac{\Phi}{y}, \quad \gamma = -\frac{\Phi}{z}. \tag{7.63}$$

[168] Expression (7.47a) then becomes

$$E = -\frac{1}{8\pi} \int \left(\mu\alpha \frac{\partial \Phi}{\partial x} + \mu\beta \frac{\partial \Phi}{\partial y} + \mu\gamma \frac{\partial \Phi}{\partial z} \right) d\omega$$

or, by integrating by parts,

$$\frac{1}{8\pi} \int \Phi \left(\frac{\partial}{\partial x} \mu\alpha + \frac{\partial}{\partial y} \mu\beta + \frac{\partial}{\partial z} \mu\gamma \right) d\omega. \tag{7.64}$$

Without any further calculation, Maxwell would have noticed that equalities (7.54b) immediately give

$$\frac{\partial}{\partial x} \mu\alpha + \frac{\partial}{\partial y} \mu\beta + \frac{\partial}{\partial z} \mu\gamma = 0, \tag{7.58b}$$

which transforms equality (7.64) into

$$E = 0.$$

The electromagnetic energy would thus be identically zero in all circumstances; such a consequence would have revealed to him that one cannot at the same time accept equalities (7.54b) and equalities (7.63). Such a contradiction does not bother Maxwell. He introduces in his calculations the quantity ρ defined by the equality

$$\frac{\partial}{\partial x} \mu\alpha + \frac{\partial}{\partial y} \mu\beta + \frac{\partial}{\partial z} \mu\gamma = 4\pi\rho; \tag{7.35b}$$

he treats this quantity ρ as if it were not identically zero and replaces equality (7.64) by the equality

$$E = \frac{1}{2} \int \Phi\rho \, d\omega. \tag{7.65}$$

---

[48] J. Clerk Maxwell, Scientific Papers, vol. I, p. 566. In fact, Maxwell called the function $-\Phi$ the *magnetic potential*.

From this expression, by a reasoning of which we have already seen several examples, Maxwell proposes to derive the law of the actions exerted between two poles of magnets.

To achieve this, Maxwell *implicitly* assumes that μ has the same value throughout all of space; the function Φ is the sum [169] two functions V_1, V_2 which are respectively the potential functions of two magnetic masses 1 and 2. The function V_1 satisfies in all space the equation

$$\Delta V_1 = 0,$$

except inside the body 1 where it satisfies the equation

$$\Delta_1 = -\frac{4\pi\rho_1}{\mu}.$$

The function V_2 satisfies, at all points in space, the equation

$$\Delta V_2 = 0,$$

except inside body 2, where it satisfies the equation

$$\Delta V_2 = -\frac{4\pi\rho_2}{\mu}.$$

The magnetic energy E can be written

$$E = -\frac{1}{8\pi} \int (V_1 + V_2)(\Delta V_1 + \Delta V_2)d\omega.$$

But according to Green's theorem,

$$\int V_1 \Delta V_2 \, d\omega = \int V_2 \Delta V_1 \, d\omega,$$

we can therefore write:

$$E = -\frac{\mu}{8\pi} \int V_1 \Delta V_1 \, d\omega - \frac{\mu}{8\pi} \int V_2 \Delta V_2 \, d\omega - \frac{\mu}{4\pi} \int V_1 \Delta V_2 \, d\omega$$

or, in virtue of the properties of the function V_2,

$$E = -\frac{\mu}{8\pi} \left(\int V_1 \Delta V_1 \, d\omega + \int V_2 \Delta V_2 \, d\omega \right) + \int V_1 \Delta \rho_2 \, d\omega.$$

Suppose that magnet 1 has fixed magnetization and position, and magnet 2 moves as a rigid solid, by [170] causing its magnetization; we can equally say that it carries with it its potential function V_2. Both integrals

$$\int V_1 \Delta V_1 \, d\omega, \quad \int V_2 \Delta V_2 \, d\omega,$$

extended to all space, clearly keep invariant values. If by dx_2, dy_2, dz_2 we mean the components of displacement of a point of the element $d\omega_2$, belonging to body 2, we will have

$$dE = \int_2 \rho_2 \left(\frac{\partial V_1}{\partial x_2} dx_2 + \frac{\partial V_1}{\partial y_2} dy_2 + \frac{\partial V_1}{\partial z_2} dz_2 \right) d\omega_2.$$

Moreover, dE is equal to the internal work accomplished in the modification in question, with a change in sign. Everything is therefore as if on each element $d\omega_2$, of body 2, body 1 exerted a force with components

$$X = -\rho_2 \frac{\partial V_1}{\partial x_2} d\omega_2, \quad Y = -\rho_2 \frac{\partial V_1}{\partial y_2} d\omega_2, \quad Z = -\rho_2 \frac{\partial V_1}{\partial z_2} d\omega_2.$$

Moreover, the analytical characteristics attributed to the function V_1 require that we have

$$V_1 = \frac{1}{\mu} \int_1 \frac{\rho_1}{r} d\omega_1.$$

The components of the force exerted by magnet 1 on an element $d\omega_2$ of magnet 2 are therefore

$$X = -\frac{1}{\mu} \rho_2 \, d\omega_2 \frac{\partial}{\partial x_2} \int_1 \frac{\rho_1}{r} d\omega_1,$$

$$Y = -\frac{1}{\mu} \rho_2 \, d\omega_2 \frac{\partial}{\partial y_2} \int_1 \frac{\rho_1}{r} d\omega_1,$$

$$Z = -\frac{1}{\mu} \rho_2 \, d\omega_2 \frac{\partial}{\partial z_2} \int_1 \frac{\rho_1}{r} d\omega_1.$$

They are the same as if two magnetic masses $m_1 = \rho_1 \, d\omega_1$ [171] and $m_2 = \rho_2 \, d\omega_2$, separated by a distance r, are repelled with a force

$$\frac{1}{\mu} \frac{m_1 m_2}{r^2}.$$

This proposition seems to be consistent with the known laws of magnetism. In reality, it is necessary to reproduce here the remark we already made: the previous theory is intimately linked to an unacceptable hypothesis; it assumes that the coefficient μ has the same value for all bodies, both for the magnets as for the medium, such as air, in which they are immersed.

7.6 The Theory of Magnetism in the TREATISE ON ELECTRICITY AND MAGNETISM

Maxwell said[49]:

> In the following Treatise I propose to describe the most important of these [electric and magnetic] phenomena, to shew how they may be subjected to measurement, and to trace the mathematical connexions of the quantities measured. Having thus obtained the data for a mathematical theory of electromagnetism, and having shewn how this theory may be applied to the calculation of phenomena, I shall endeavour to place in as clear a light as I can the relations between the mathematical form of this theory and that of the fundamental science of Dynamics, in order that we may be in some degree prepared to determine the kind of dynamical phenomena among which we are to look for illustrations or explanations of the electromagnetic phenomena.

The object of the work thus being clearly defined, the following problem must play an essential role:

From basic laws of electricity and magnetism, to derive the expression of the electrostatic energy and electromagnetic energy; to show that these two energies can [172] be put in the form that the memoir: *On Physical Lines of Force* has attributed to the potential energy and to the live force[50] of the medium whose mechanical deformations imitate or explain the electromagnetic phenomena.

We have already seen[51] how the part of this program which concerns electrostatic energy is realized. Now let us examine the determination of electromagnetic energy.

Maxwell arrives at the expression of this energy by two different methods; one of these methods involves the laws of electromagnetism, while the other, restricted to systems that contain no currents, relies exclusively on the theory of magnetism.

The *Treatise on Electricity and Magnetism*, indeed, presents a complete theory of magnetism. This theory forms the third part of the book.

The theory of magnetism Maxwell presents is the classical theory created by the work of Poisson, F. E. Neumann, G. Kirchhoff, and W. Thomson, the theory whose key proposals we have summarized previously.[52] He considers, in particular, the intensity of magnetization, defined as we defined it in the passage quoted.

Components A, B, C of this intensity of magnetization are used, by Maxwell as by Poisson,[53] to define the magnetic potential function by the formula

$$V = \int \left(A_1 \frac{\partial \frac{1}{r}}{\partial x_1} + B_1 \frac{\partial \frac{1}{r}}{\partial y_1} + C_1 \frac{\partial \frac{1}{r}}{\partial z_1} \right) d\omega_1. \tag{7.66}$$

---

[49] J. Clerk Maxwell, *Treatise on Electricity and Magnetism*, Preface of the 1st edition; t. 1, p. IX de la traduction française [pp. v–vi of the English original].

[50] [*Force vive* or mv^2, related to the kinetic energy $mv^2/2$].

[51] 1st Part, Chap. 5, Sect. 5.2.

[52] 1st Part, Chap. 2, Sect. 2.1.

[53] 1st Part, equality (2.1).—J. Clerk Maxwell, *Treatise on Electricity and Magnetism*, trad. française, t. II, p. 10 [9], equality (8).

Components α, β, γ of the field are linked to this function by the relations

$$\alpha = -\frac{\partial V}{\partial x}, \quad \beta = -\frac{\partial V}{\partial y}, \quad \gamma = -\frac{\partial V}{\partial z}. \tag{7.67}$$

[173] This potential function can also be expressed by the means of two solid and surface densities, ρ and σ, of the fictitious magnetic fluid by the equality

$$V = \int \frac{\rho_1}{r} d\omega_1 + \int \frac{\sigma_1}{r} dS_1,$$

and these densities are related to the components of magnetization by equalities[54]

$$\frac{\partial A}{\partial x} + \frac{\partial B}{\partial y} + \frac{\partial C}{\partial z} = -\rho, \tag{7.68}$$

$$A \cos(N_i, x) + B \cos(N_i, x) + C \cos(N_i, x) = -\sigma, \tag{7.69}$$

already given by Poisson.

In comparison to the intensity of magnetization, but without confusing it with it, as he seems to have done in his early writings, Maxwell considers[55] the *magnetic induction*. The components A, B, C of this quantity are defined by the equalities

$$\begin{cases} A = \alpha + 4\pi A, \\ B = \beta + 4\pi B, \\ C = \gamma + 4\pi C, \end{cases} \tag{7.70}$$

which equalities (7.67) also allow us to write

$$\begin{cases} A = 4\pi A - \dfrac{\partial V}{\partial x}, \\ B = 4\pi B - \dfrac{\partial V}{\partial y}, \\ C = 4\pi C - \dfrac{\partial V}{\partial z}, \end{cases} \tag{7.71}$$

In restoring the magnetic induction to its proper meaning, Maxwell drops the relationship, under two different and [174] incompatible forms, that he intended to establish between the components of the magnetic induction and the density of the fictitious magnetic material. Hence, he implicitly denies all reasonings, so essential in his previous writings, which invoked this relationship.

---

[54] 1st Part, equalities (2.2) and (2.3). — J. Clerk Maxwell, *loc. cit.*, p. 11 [10].

[55] J. Clerk Maxwell, *loc. cit.*, nº 400, p. 28 [24].

At any point of a continuous medium, we have[56]

$$\frac{\partial^2 V}{\partial x^2} + \frac{\partial^2 V}{\partial y^2} + \frac{\partial^2 V}{\partial z^2} = 4\pi \left(\frac{\partial A}{\partial x} + \frac{\partial B}{\partial x} + \frac{\partial C}{\partial x} \right).$$

In virtue of equalities (7.71), this equality becomes[57]

$$\frac{\partial A}{\partial x} + \frac{\partial B}{\partial x} + \frac{\partial C}{\partial x} = 0. \qquad (7.72)$$

From this last equality, it follows that one can find three functions F, G, H, such that we have

$$\begin{cases} \dfrac{\partial H}{\partial y} - \dfrac{\partial G}{\partial z} = -A, \\[2mm] \dfrac{\partial F}{\partial z} - \dfrac{\partial H}{\partial x} = -B, \\[2mm] \dfrac{\partial G}{\partial z} - \dfrac{\partial F}{\partial y} = -C. \end{cases} \qquad (7.73a)$$

Maxwell writes these equations[58] and calls quantity, of which F, G, H are the components, the vector potential of magnetic induction. In this regard, we need to repeat the observation that we have already made regarding equalities (7.54b): equalities (7.73a) are not sufficient to determine the functions F, G, H, as long as the value of the sum

$$\frac{\partial F}{\partial x} + \frac{\partial F}{\partial x} + \frac{\partial F}{\partial x}$$

is left indeterminate.

[175] In a perfectly soft body where the magnetizing function is reduced to a coefficient k independent of the intensity of magnetization, we have

$$A = k\alpha, \quad B = k\beta, \quad C = k\gamma. \qquad (7.74)$$

Equalities (7.70) then become

$$A = \frac{1 + 4\pi k}{k} A, B = \frac{1 + 4\pi k}{k} B, C = \frac{1 + 4\pi k}{k} C.$$

Among the components of magnetization and magnetic induction, we find relations (7.33). If we set

$$\mu = 1 + 4\pi k, \qquad (7.75)$$

---

[56] 1st Part, equality (2.4).

[57] J. Clerk Maxwell, *loc. cit.*, p. 57 [50], equality (17).

[58] J. Clerk Maxwell, *loc. cit.*, n° 405, p. 32 [28], equalities (21). In the *Treatise* of Maxwell, the sign of the second member is changed as the result of a different choice of coordinate axes.

equalities (7.70) and (7.74) give the equalities[59]

$$A = \mu\alpha, \quad B = \mu\beta, \quad C = \mu\gamma, \tag{7.76a}$$

which, in the earlier writings of Maxwell, were used to define the magnetic induction.

We come to the determination of the magnetic energy.

Magnet 1, of which (x_1, y_1, z_1) is a point and $d\omega_1$ an element, lies in the presence of another magnet, whose magnetic potential function is V_2. These two magnets are rigid solids, and to each of their elements an intensity of magnetization is invariably linked; while magnet 2 remains stationary, magnet 1 moves. The actions of magnet 2 on magnet 1 produce a certain work; according to the classical [176] doctrines of magnetism, this work is equal to the decrease suffered by the quantity

$$W = \int \left(A_1 \frac{\partial \frac{1}{r}}{\partial x_1} + B_1 \frac{\partial \frac{1}{r}}{\partial y_1} + C_1 \frac{\partial \frac{1}{r}}{\partial z_1} \right) d\omega_1.$$

Maxwell demonstrates this proposition,[60] which is universally accepted.

To take this proposition as starting point and conclude that the energy of any system of magnetic bodies is given by the expression

$$E = \frac{1}{2} \int \left(A \frac{\partial V}{\partial x} + B \frac{\partial V}{\partial y} + C \frac{\partial V}{\partial z} \right) d\omega, \tag{7.77}$$

where V is the magnetic potential function of the entire system and where the integral extends over the entire system, is obviously to make an hypothesis. This hypothesis and the recent progress of thermodynamics show that it is not justified, but it must have seemed natural at the time Maxwell wrote; also, Maxwell adopts it.[61]

Therefore, a classical transformation allows us to write

$$E = \frac{1}{8\pi} \int \left[\left(\frac{\partial V}{\partial x} \right)^2 + \left(\frac{\partial V}{\partial y} \right)^2 + \left(\frac{\partial V}{\partial z} \right)^2 \right] d\omega$$

or, in virtue of equalities (7.67),

$$E = \frac{1}{8} \int (\alpha^2 + \beta^2 + \gamma^2) d\omega. \tag{7.78}$$

This is the expression of the magnetic energy at which Maxwell arrives.[62]

---

[59] J. Clerk Maxwell, *loc. cit.*, p. 57 [50], equalities (16).
[60] J. Clerk Maxwell, *loc. cit.*, p. 18 [15], equality (3).
[61] J. Clerk Maxwell, *loc. cit.*, p. 304 [247], expression (6).
[62] J. Clerk Maxwell, *loc. cit.*, p. 305 [247], equality (11).

This expression does not coincide with the expression (7.47a) that he wanted to find; the factor μ is lacking in the integrand. [177] To find the expression of electromagnetic energy he wishes to reach, Maxwell must appeal to the theory of electromagnetism.

7.7 The Theory of Electromagnetism in the TREATISE ON ELECTRICITY AND MAGNETISM

We briefly summarize the theory of electromagnetism, such as Maxwell presents it in his *Treatise*.

He first introduces a vector, with components F, G, H, to which, in a stationary system of variable electrical state, the components E'_x, E'_y, E'_z of the electromotive induction field should be linked by the equalities[63]

$$E'_x = -\frac{\partial F}{\partial t}, \quad E'_y = -\frac{\partial G}{\partial t}, \quad E'_z = -\frac{\partial H}{\partial t}. \tag{7.79}$$

This vector is therefore what he had called, in his previous writings, the *electrotonic state* or *electromagnetic moment*; he now names it the *quantity of electrokinetic movement*[64]; then, immediately, he issues this assertion[65]:

[This vector] is identical with the quantity which we investigated…under the name of the vector-potential of magnetic induction.

In support of this assertion, Maxwell outlines a *prima facie* case.[66] Expressions (7.79) of the electromotive induction field, applied to a closed stationary wire, give the following expression for the total electromotive force of induction which acts in this wire:

$$-\int \left(\frac{\partial F}{\partial t}\frac{dx}{ds} + \frac{\partial G}{\partial t}\frac{dy}{ds} + \frac{\partial H}{\partial t}\frac{\partial z}{\partial s} \right) ds.$$

In this expression, the integral extends over all the linear elements ds into which the wire can be divided. [178]

We take the wire to outline an area of which dS is an element and let N be a normal to the element dS, directed in a suitable sense. We know that the previous expression can be written

---

[63] J. Clerk Maxwell, *loc. cit.*, p. 267 [214] and p. 274 [221], equalities (B).

[64] J. Clerk Maxwell, *loc. cit.*, p. 267 [214].

[65] J. Clerk Maxwell, *loc. cit.*, p 267 [214].

[66] J. Clerk Maxwell, *loc. cit.*, p. 268 [215], n° 592.

$$\int \left[\frac{\partial}{\partial t} \left(\frac{\partial H}{\partial y} - \frac{\partial G}{\partial z} \right) \cos (N, x) + \frac{\partial}{\partial t} \left(\frac{\partial F}{\partial z} - \frac{\partial H}{\partial x} \right) \cos (N, y) \right.$$

$$\left. + \frac{\partial}{\partial t} \left(\frac{\partial G}{\partial x} - \frac{\partial F}{\partial y} \right) \cos (N, z) \right] dS,$$

the integral extending over the considered area.

But, secondly, *if the wire is placed in a non-magnetic medium*, it is known from Faraday that the electromotive force is related to the variation of the magnetic field by the formula

$$- \int \left[\frac{\partial \alpha}{\partial t} \cos (N, x) + \frac{\partial \beta}{\partial t} \cos (N, y) + \frac{\partial \gamma}{\partial t} \right] dS.$$

Equalities (7.79) will therefore agree with the laws of induction in a closed circuit, in a non-magnetic medium, if we have

$$\frac{\partial H}{\partial y} - \frac{\partial G}{\partial z} = -\alpha, \quad \frac{\partial F}{\partial z} - \frac{\partial H}{\partial x} = -\beta, \quad \frac{\partial G}{\partial x} - \frac{\partial F}{\partial y} = -\gamma. \tag{7.80}$$

Equalities (7.80) can be viewed as particular instances of equalities

$$\begin{cases} \dfrac{\partial H}{\partial y} - \dfrac{\partial G}{\partial z} = -(\alpha + 4\pi A) - A, \\[2mm] \dfrac{\partial F}{\partial z} - \dfrac{\partial H}{\partial x} = -(\beta + 4\pi B) - B, \\[2mm] \dfrac{\partial G}{\partial z} - \dfrac{\partial F}{\partial y} = -(\gamma + 4\pi C) - C. \end{cases} \tag{7.73b}$$

They do not justify them, but they make the hypothesis that Maxwell made acceptable, by adopting[67] equalities (7.73b). [179]

When the magnetic medium is perfectly soft and where the magnetizing function of this medium is reduced to a coefficient independent of the intensity of magnetization,[68] we have

$$A = \mu\alpha, \quad B = \mu\beta, \quad C = \mu\dot{\gamma}, \tag{7.76b}$$

[67] J. Clerk Maxwell, *loc. cit.*, p. 268 [215], equalities (A).
[68] J. Clerk Maxwell, *loc. cit.*, p. 289 [233], n° 614.

and equalities (7.73b) take the form

$$
\begin{cases}
\dfrac{\partial H}{\partial y} - \dfrac{\partial G}{\partial z} = -\mu\alpha, \\[2mm]
\dfrac{\partial F}{\partial z} - \dfrac{\partial H}{\partial x} = -\mu\beta, \\[2mm]
\dfrac{\partial G}{\partial x} - \dfrac{\partial F}{\partial y} = -\mu\gamma,
\end{cases}
\tag{7.54b}
$$

already given in the memoir: *A Dynamical Theory of the Electromagnetic Field.*
Along with the equalities[69]

$$
\begin{cases}
\dfrac{\partial \gamma}{\partial y} - \dfrac{\partial \beta}{\partial z} = -4\pi(u + \bar{u}), \\[2mm]
\dfrac{\partial \alpha}{\partial z} - \dfrac{\partial \gamma}{\partial x} = -4\pi(v + \bar{v}), \\[2mm]
\dfrac{\partial \beta}{\partial x} - \dfrac{\partial \alpha}{\partial y} = -4\pi(w + \bar{w}),
\end{cases}
\tag{7.5}
$$

equations (7.54b) form the now famous group of Maxwell's six equations.

The functions F, G, H which are included in equations (7.54b) are not fully defined. Twice already we have made this remark; to determine them, we need to know the value of the quantity[70]

$$
\frac{\partial F}{\partial x} + \frac{\partial G}{\partial y} + \frac{\partial H}{\partial z} = J.
\tag{7.55b}
$$

[180] Now, this quantity has an unknown value, which creates an obstacle in the following calculation [71]:

Equalities (7.54b) and (7.5) easily give the relations

$$
\Delta F = \frac{\partial J}{\partial x} - 4\pi\mu(u + \bar{u}),
$$

$$
\Delta G = \frac{\partial J}{\partial y} - 4\pi\mu(v + \bar{v}),
$$

$$
\Delta H = \frac{\partial J}{\partial z} - 4\pi\mu(w + \bar{w}).
$$

---

[69] J. Clerk Maxwell, *loc. cit.*, p. 286 [231], equalities (E) and p. 290 [234].

[70] J. Clerk Maxwell, *loc. cit.*, p. 290 [235], equality (2).

[71] J. Clerk Maxwell, *loc. cit.*, p. 290 [234], n° 616.—This calculation was already found almost verbatim in the memoir: *A Dynamical Theory of the Electromagnetic Field* (J. Clerk Maxwell, SCIENTIFIC PAPERS, vol. I, p. 581).

So if we put

$$\begin{cases} F' = \displaystyle\int \frac{\mu_1(u_1 + \bar{u}_1)}{r} d\omega_1, \\[2mm] C' = \displaystyle\int \frac{\mu_1(v_1 + \bar{v}_1)}{r} d\omega_1, \\[2mm] H' = \displaystyle\int \frac{\mu_1(w_1 + \bar{w}_1)}{r} d\omega_1, \end{cases} \tag{7.81}$$

$$\chi = -\frac{1}{4\pi} \int \frac{J_1}{r} d\omega_1, \tag{7.82}$$

formulas where the integrations extend over all space, we will have

$$\begin{cases} F = F' + \dfrac{\partial \chi}{\partial x}, \\[2mm] G = G' + \dfrac{\partial \chi}{\partial y}, \\[2mm] H = H' + \dfrac{\partial \chi}{\partial z}. \end{cases} \tag{7.83}$$

[181] Maxwell adds [72]:

The quantity χ disappears from the equations (A) [(7.54b)], and it is not related to any physical phenomenon. If we suppose it to be zero everywhere, J will also be zero everywhere, and equations (5) [(7.81)], omitting the accents, will give the true values of the components of...[*the vector potential*].

The quantity χ, certainly, disappears from equalities (7.54b); but it appears in equalities (7.79). Is it thus so obvious that it has no influence on any physical phenomenon? Without a doubt, the total electromotive force that acts in a closed circuit

$$-\int \left(\frac{\partial F}{\partial t} \frac{\partial x}{\partial s} + \frac{\partial G}{\partial t} \frac{\partial y}{\partial s} + \frac{\partial H}{\partial t} \frac{\partial z}{\partial s} \right) ds$$

can also be written

$$-\int \left(\frac{\partial F'}{\partial t} \frac{\partial x}{\partial s} + \frac{\partial G'}{\partial t} \frac{\partial y}{\partial s} + \frac{\partial H'}{\partial t} \frac{\partial z}{\partial s} \right) ds,$$

and its value will be independent of the determination attributed to the function χ; but it does not follow that this would not interfere with any discussion of physics. Asserting this would accuse Maxwell of an absurdity in a passage that he wrote[73] in all what preceded.

---

[72] J. Clerk Maxwell, *loc. cit.*, p. 291 [235].
[73] J. Clerk Maxwell, *loc. cit.*, p. 274 [221].

The electromotive force which acts in a closed circuit is given by the expression

$$-\int \left(\frac{\partial F}{\partial t}\frac{\partial x}{\partial s} + \frac{\partial G}{\partial t}\frac{\partial y}{\partial s} + \frac{\partial H}{\partial t}\frac{\partial z}{\partial s} \right) ds.$$

Maxwell concluded that the electromotive field has components at each point

$$E_x = -\frac{\partial \Psi}{\partial x} - \frac{\partial F}{\partial t}, \quad E_x = -\frac{\partial \Psi}{\partial y} - \frac{\partial G}{\partial t}, \quad E_x = -\frac{\partial \Psi}{\partial z} - \frac{\partial H}{\partial t}; \qquad (7.84)$$

and then he adds[74]:

> The terms involving the new quantity Ψ are introduced for the sake of giving generality to the expressions for...[182] [E_x, E_y, E_z]. They disappear from the integral when extended round the closed circuit. The quantity Ψ is therefore indeterminate as far as regards the problem now before us, in which the total electromotive force round the circuit is to be determined. We shall find, however, that when we know all the circumstances of the problem, we can assign a definite value to Ψ, and that it represents, according to a certain definition, the electric potential at the point (x, y, z).

If the function Ψ plays a role in the analysis of some problems of electricity, why does not the function χ play any?

These two groups of equations, (7.54b) and (7.5), will provide for Maxwell the expression of electromagnetic energy that he wants to obtain, by a calculation that is similar to what is given in the memoir: *A Dynamical Theory of the Electromagnetic Field* and what we presented in Sect. 7.5.

This energy, according to Maxwell's first expression, is[75]

$$E = \frac{1}{2} \int [F(u + \overline{u}) + G(v + \overline{v}) + H(w + \overline{w})] \, d\omega.$$

Equalities (7.5) transform it into

$$E = -\frac{1}{8\pi} \int \left[\left(\frac{\partial \gamma}{\partial y} - \frac{\partial \beta}{\partial z} \right) F + \left(\frac{\partial \alpha}{\partial z} - \frac{\partial \gamma}{\partial x} \right) G + \left(\frac{\partial \beta}{\partial x} - \frac{\partial \alpha}{\partial y} \right) H \right] d\omega.$$

Integration by parts gives

$$E = -\frac{1}{8\pi} \int \left[\left(\frac{\partial H}{\partial y} - \frac{\partial G}{\partial z} \right) \alpha + \left(\frac{\partial F}{\partial z} - \frac{\partial H}{\partial x} \right) \beta + \left(\frac{\partial G}{\partial x} - \frac{\partial F}{\partial y} \right) \gamma \right] d\omega.$$

or, in virtue of equalities (7.73b),

$$E = \frac{1}{8\pi} \int (A\alpha + B\beta + C\gamma) \, d\omega. \qquad (7.85)$$

---

[74][*ibid.*].

[75]J. Clerk Maxwell, *loc. cit.*, p. 305 [248], n° 634 to 636.

In the case of the system containing only a perfectly soft [183] magnetic body, equalities (7.76b) transform equality (7.85) into

$$E = \frac{1}{8\pi} \int \mu(\alpha^2 + \beta^2 + \gamma^2)d\omega. \tag{7.47b}$$

Thus we find by an electrodynamic method the expression of electromagnetic energy what the memoir: *On Physical Lines of Force* had obtained by the use of mechanical hypotheses.

Both expressions (7.78) and (7.47b) of the electromagnetic energy do not agree; this disagreement does not escape Maxwell and only embarrasses him. First, he says,[76] speaking of magnetic energy taken in the form (7.78), that "This part of the energy, however, will be included in the kinetic energy in the form in which we shall presently obtain it," i.e. in the form (7.47b); but then he recognizes[77] that the expression obtained for electromagnetic energy, together with the hypothesis that such energy represents the live force, cannot agree with the usual theory of magnetism:

> This mode of explaining magnetism requires us also to abandon the method followed in Part III, in which we regarded the magnet as a continuous and homogeneous body, the minutest part of which has magnetic properties of the same kind as the whole. We must now regard a magnet as containing a finite, though very great, number of electric circuits...

In writing his treatise, Maxwell proposed to take as a point of departure the well-established laws of electricity and magnetism and translate them by equations whose form would leave transparent the relationships between these laws and the principles of dynamics; but the reality remains very far removed from that promise, and rather than give up a mechanical interpretation in which he above all holds, Maxwell likes better to dispense with one of the most perfect branches of sound physics, the theory of magnetism; thus we had seen him in our First Part leaving, for the more adventurous hypotheses, the formerly most well-established electric doctrine: electrostatics. [184]

Maxwell's electrodynamics proceeds by following the unusual method that we have already analyzed in studying electrostatics. Under the influence of hypotheses that remain in his mind vague and imprecise, Maxwell drafts a theory that he does not complete, from which he does not even bother to remove the contradictions. Then, he constantly modifies this theory, imposing fundamental changes that he does not report to his reader; and he makes vain efforts to fix his fleeting and elusive thinking. At the moment he thinks he succeeds, he sees the same parts of the doctrine which relate to the best-studied phenomena vanish.

This strange and disconcerting method is, however, what led Maxwell to the electromagnetic theory of light. [185]

---

[76] J. Clerk Maxwell, *loc. cit.*, p. 304 [247].

[77] J. Clerk Maxwell, *loc. cit.*, p. 309 [251].

Chapter 8
The Electromagnetic Theory of Light

8.1 The Speed of Light and the Propagation of Electrical Actions: The Research of W. Weber and G. Kirchhoff

It is to Wilhelm Weber that one must turn to find the first mention, in his study of electrical phenomena, of the number that measures the speed of the propagation of light in a vacuum.

The first study published[1] in 1846 by W. Weber, under the title: *Elektrodynamische Maassbestimmungen*,[2] contained an appendix entitled:

Ueber die Zusammenhang der elektrostatischen und der elektrodynamischen Erscheinungen nebst Anwendung auf die elektrodynamischen Maassbestimmungen.

This appendix contained the famous law of Weber.

A wire carrying an electric current is actually the seat of two currents of opposite direction: one, headed in the direction of the current, carries positive electricity; the other headed in the opposite direction, carries negative electricity. When the current is uniform, these two currents have an equal flow.

On the other hand, the law of mutual action of two electrical charges expressed by Coulomb is an incomplete law; it applies only to charges that are in relative rest. If two electrical charges e, e' are separated by a distance r which varies with the [186] time t, these two charges repel by force whose expression is

$$\frac{ee'}{r^2}\left[1 - \frac{a^2}{16}\left(\frac{dr}{dt}\right)^2 + \frac{a^2}{8}r\frac{d^2r}{dt^2}\right].$$

Applied to the computation of electrodynamic actions, this law gives back the elementary law of Ampère; applied to the phenomena of induction, it formulates the mathematical law.

---

[1] W. Weber, *Elektrodynamische Maassbestimmungen*, Leipzig, 1846.
[2] [English translation: Weber (2007)].

© Springer International Publishing Switzerland 2015
P.M.M. Duhem, *The Electric Theories of J. Clerk Maxwell*,
Boston Studies in the Philosophy and History of Science 314,
DOI 10.1007/978-3-319-18515-6_8

The constant a figures in each of these laws. Let us see, in particular, how it appears in Ampère's law.

Two elements of uniform current ds, ds' are present. In the first, the current of positive electricity and the current of negative electricity have a common value, i. In the second, these two currents have a common value, i'. The angle of the two elements is ε; r is the distance that separates them; and θ, θ' are the angles that these elements make with the line that goes from a point of the element ds to a point of the element ds'. These two elements repel with a force

$$a^2 \frac{i\, ds\, i'\, ds'}{r^2} \left(\cos \varepsilon - \frac{3}{2} \cos \theta \cos \theta' \right).$$

The *intensities* J, J' of the two currents are related to the partial currents i, i' by the relations

$$J = 2i, \quad J' = 2i'.$$

The previous force can then also be written:

$$-\frac{a^2}{4} \frac{J\, ds\, J'\, ds'}{r^2} \left(\cos \varepsilon - \frac{3}{2} \cos \theta \cos \theta' \right).$$

Today, one usually writes this formula in the following manner:

$$-2A^2 \frac{J\, dS\, J'\, ds'}{r^2} \left(\cos \varepsilon - \frac{3}{2} \cos \theta \cos \theta' \right),$$

A being the *fundamental constant of electromagnetic actions* [187] *evaluated in electrostatic units*. Weber's constant a^2 is, one can see, related to the constant A^2 by the relation

$$A^2 = \frac{a^2}{8}. \tag{8.1}$$

W. Weber, moreover, soon changed the form of his law, writing

$$\frac{ee'}{r^2} \left[1 - \frac{1}{c^2} \left(\frac{dr}{dt} \right)^2 + \frac{2}{c^2} r \frac{d^2 r}{dt^2} \right].$$

The new constant c^2, thus introduced, was linked to the constant a by the equality

$$\frac{1}{c^2} = \frac{a^2}{16},$$

and, therefore, in virtue of equality (8.1), it is linked to the constant A^2 by the equality

$$A^2 = \frac{2}{c^2}. \tag{8.2}$$

It is clear that c is a quantity of the same sort as a speed.

We imagine that the speed with which both charges e, e' approach or move away from each other, the speed whose absolute value is that of $\frac{dr}{dt}$, be a uniform speed; $\frac{d^2r}{dt^2}$ will be equal to 0 and the two charges will repel each other with a force

$$\frac{ee'}{r^2}\left[1 - \frac{1}{c^2}\left(\frac{dr}{dt}\right)^2\right].$$

If we have

$$\left(\frac{dr}{dt}\right)^2 = c^2,$$

these two forces will cancel themselves out; the electrodynamic force $-\frac{1}{c^2}\frac{ee'}{r^2}\left(\frac{dr}{dt}\right)^2$ will balance the electrostatic force $\frac{ee'}{r^2}$. [188]

W. Weber and R. Kohlrausch, in a memoir that has remained classic,[3] experimentally determined the value of this constant c. They found that this value, evaluated *in millimeters per second*, was:

$$c = 439\,450 \times 10^6.$$

Following this outcome of the review, they simply write:

> This determination of the constant c thus proves that two electric masses must move with a very high velocity relative to the other, if one wants the *electrodynamic force* to eliminate the *electrostatic force*; namely, with a speed of 439 million meters [per second] or 59 320 miles[4] per second, which significantly surpasses the speed of light.[5]

---

[3]R. Kohlrausch and W. Weber, *Elektrodynamische Maassbestimmungen, insbesondere Zurückführung der Stromintensitäts-Messungen auf mechanische Maass*, Leipzig, 1856. [English translation: Weber and Kohlrauch (2003)].

[4][≈3–6 English miles (Oxford English Dictionary, 2014)].

[5]Weber (1893, p. 652):

> Aus dieser Bestimmung der *Konstanten c* ersieht man also, dass zweielektrische Massen mit sehr grosser Geschwindigkeit gegen einander bewegt werden müssen, wenn die *elektrodynamische* Kraft die *elektrostatische* aufheben soll, nämlich mit einer Geschwindigkeit von 439 Millionen Meter oder 59 320 Meilen in der Sekunde, welche die Geschwindigkeit des Lichts bedeutend übertrifft.

The following year, G. Kirchhoff[6] proposed to deduce from Weber's theory the laws according to which electrodynamic induction propagates in a wire.

He remarked that the resistance of the wire was included in the resulting equations, but divided by a constant factor whose numeric value is extremely large; so that in a copper wire of a few meters in length and a few millimeters in radius, the laws of the variation of electric current were essentially the same as if the wire had no resistance. In this limit, where the wire is assumed to be without resistance, the intensity J of the electric current that runs through a closed conductor is expressed, at the moment t, by the following formula:

$$J = -\frac{c}{4\sqrt{2}}e^{-ht}\left[f\left(s + \frac{c}{\sqrt{2}}t\right) + f\left(s - \frac{c}{\sqrt{2}}t\right)\right],$$

s being the length of the wire from a given origin to the point under consideration, h a constant, and f an arbitrary function. [189]

This current can be regarded as the result of the superposition of two other currents of respective intensities

$$J' = -\frac{c}{4\sqrt{2}}e^{-ht}\left[f\left(s + \frac{c}{\sqrt{2}}t\right)\right],$$

$$J'' = -\frac{c}{4\sqrt{2}}e^{-ht}\left[f\left(s - \frac{c}{\sqrt{2}}t\right)\right],$$

or of two damped waves that propagate in a contrary sense with a speed $\frac{c}{\sqrt{2}}$.

Kirchhoff said:

> The velocity of propagation of an electric wave is here equal to $\frac{c}{\sqrt{2}}$; it is therefore independent of the cross-section of the wire, of its conductivity, and, finally, of the electric density; its value is 41 950 miles per second; it is thus very near the speed of light in empty space.[7]

The analysis of the movement of electricity through a wire, which led G. Kirchhoff to this remarkable result, was extended shortly thereafter[8] by the same author to conductors whose three dimensions are finite.

The result obtained by G. Kirchhoff could not fail to strike Weber. He undertook to submit the oscillations of a varying electric current in a conductor to an in-depth

---

[6]G. Kirchhoff, *Ueber die Bewegung der Elektricität in Drähten* (POGGENDORFF'S ANNALEN), Bd. C, 1857. [English translation: Kirchhoff (1857a)].

[7]Kirchhoff (1857b, p. 209–210):

> Die Fortpflanzungsgeschwindigkeit einer elektrischen Welle hat sich hier = $\frac{c}{\sqrt{2}}$ ergeben, also als unabhängig sowohl von dem Querschnitt, als von der Leitungsfähigkeit des Drahtes, als endlich von der Dichtigkeit der Elektrizität; ihr Wert ist der von 41 950 Meilen in einer Sekunde, also sehr nahe gleich der Geschwindigkeit des Lichtes im leeren Raume.

[8]G. Kirchhoff, *Ueber die Bewegung der Elektricität in Leitern* (POGGENDORFF'S ANNALEN), Bd. CII, 1857. [English translation: Graneau and Assis (1994)].

theoretical and experimental study.[9] This study confirms the researches of Kirchhoff. Subject to certain hypotheses, among which is the low resistance of the wire, it is recognized that

$\frac{c}{\sqrt{2}}$ is therefore the limiting value to which all the propagation velocities approach, and, for the given value of c,

$$c = 439\,450 \times 10^6 \frac{\text{millimeters}}{\text{second}},$$

[190] this limit has the value

$$\frac{c}{\sqrt{2}} = 310\,740 \times 10^6 \frac{\text{millimeters}}{\text{second}},$$

which is a speed of 41 950 miles per second.[10]

G. KIRCHHOFF has already found this expression for the propagation speed of electric waves, and he remarked "that it is independent of the cross-section of the wire, its conductivity, and electric density; its value, which is 41 950 miles per second, is very close to the speed of light in a vacuum." If this close match between electric wave propagation speed and the speed of light could be regarded as an indication of an inner relationship between the two doctrines, it will deserve the greatest interest, because finding such a relationship is of great importance. But the true meaning that this speed has regarding electricity must, above all, be considered; and this meaning does not appear to favor great expectations.[11]

---

[9]Wilhem Weber, *Elektrodynamische Maassbestimmungen, insbesondere über elektrische Schwingungen*, Leipzig, 1864.

[10]Weber (1894, p. 157):

...ist daher $c/\sqrt{2}$ der gesuchte Grenzwerth, dem sich alle Fortpflanzungsgeschwindigkeiten nähern, und dieser Grenzwerth ist, für den gegebenen Werth $c = 439\,450 \times 10^6$ Millimeter/Sekunde,

$$\frac{c}{\sqrt{2}} = 310\,740 \times 10^6 \frac{\text{Millimeter}}{\text{Sekunde}},$$

d.i. eine Geschwindigkeit von 41 950 Meilen in der Sekunde.

[11]Weber (1894, *loc.cit.*):

Diese Geschwindigkeit hat schon KIRCHHOFF für die Fortpflanzung elektrischer Wellen gefunden und bemerkt: "dass sie sowohl unabhängig von dem Querschnitt, als auch von der Leitungsfähigkeit des Drahts, als auch endlich von der Dichtigkeit der Elektricität wäre; auch dass ihr Werth von 41 950 Meilen in einer Sekunde sehr nahe dem der Geschwindigkeit des Lichts im leeren Raume gleichkommt". Könnte diese nahe Uebereinstimmung der Fortpflanzungsgeschwindigkeit elektrischer Wellen mit der des Lichts als eine Andeutung eines inneren Zusammenhangs beider Lehren angesehen werden, so würde sie bei der grossen Wichtigkeit, welche die Erforschung eines solchen Zusammenhangs hat, das grösste Interesse in Anspruch nehmen. Es leuchtet aber ein, dass dabei vor Allem die wahre Bedeutung, die in Beziehung auf die Elektricität jener Geschwindigkeit zukommt, in Betracht gezogen werden muss, welche nicht der Art zu sein scheint, dass sich grosse Erwartungen daran knüpfen liessen.

As we have shown above, so that the true speed of propagation approaches this limit which coincides with the speed of light, it is necessary not only that the wire be very thin compared to its length, but also that this long and thin wire have a very low resistance. It is quite clear that the actual speed will only come near this limiting value and that, very frequently, it will be quite far from it.[12]

8.2 The Speed of Light and the Propagation of Electrical Actions: The Research of B. Riemann, C. Neumann, and L. Lorenz

The, at least approximate, equality

$$A^2 = \frac{1}{V^2},\qquad\qquad(8.3a)$$

where V refers to the speed of light in a vacuum, is no less a consequence of the experiments of Weber and Kohlrausch, [191] and, despite the approximations to which the proposition demonstrated by G. Kirchhoff was submitted, this equality was too striking so that we only see in it the mark of an intimate relationship between light and electricity. From this moment onward, physicists tried to introduce into electrical theories the idea of a propagation that would occur throughout space at the speed of light.

On 10 February 1858, Bernhard Riemann read to the Society of Sciences of Göttingen a note entitled: *Ein Beitrag zur Elektrodynamik*; this note was published[13] after the death of the illustrious mathematician.

The point of departure adopted by Riemann is the following.

Suppose a point M carries, at time t, an electrical charge that can vary with t, $q(t)$. It is generally assumed that at a point M', of which r is the distance to the point M, this electric charge produces a potential function whose value, at the same time t, is $\frac{q(t)}{r}$. At the instant t, the potential function at the point M' is

$$V' = \sum \frac{q(t)}{r}.$$

---

[12]Weber (1894, p. 157-8):

Denn die Annäherung der wahren Fortpflanzungsgeschwindigkeit an jenen Grenzwerth, der mit der Geschwindigkeit des Lichts übereinstimmt, setzt, wie eben gezeigt worden, nicht blos einen im Vergleich zu seiner Länge sehr dünnen Leitungsdraht voraus, sondern auch, dass dieser lange und dünne Leitungsdraht einen sehr kleinen Widerstand besitze. Es leuchtet hieraus ein, dass grössere Annäherung an jenen Grenzwerth nur selten, grössere Abweichungen davon sehr häufig vorkommen werden.

[13]Bernhard Riemann, *Ein Beitrag zur Elektrodynamik*, POGGENDORFF'S ANNALEN, Bd. CXXXI.— *Bernhard Riemann's gesammelte mathematische Werke*, p. 270; 1876.

Riemann admits that at the moment t, the potential function caused in M' by the charge of the point M is $\frac{1}{r}q\left(t - \frac{r}{a}\right)$, a being a positive constant; the potential function in M' at time t is

$$V' = \sum \frac{q\left(t - \frac{r}{a}\right)}{r}.$$

One can clearly articulate this hypothesis by saying that the electrostatic potential function, instead of instantly propagating in space, as one usually admits, is propagating through it with the finite speed a. [192]

Now, from this hypothesis Bernhard Riemann deducted, for the mutual electrodynamic potential of both systems, a formula that coincides with what W. Weber gave, provided that we take

$$a = \frac{c}{\sqrt{2}}.$$

According to the determination of Weber and Kohlrausch, it is

$$c = 439\,450 \times 10^6 \frac{\text{millimeters}}{\text{second}}.$$

The result is that a is equal to 41 949 geographical miles per second, while the calculations of Busch, from Bradley's aberration observations, gives the speed of light to be 41 994 miles [per second], and Fizeau found it to be, by direct measurement, 41 882 miles [per second].[14]

Riemann could thus summarize his contribution to electrodynamics as follows:

I found that one could explain the electrodynamic actions of electrical currents assuming that the action of one electrical mass on another does not occur instantly, but propagates with a constant speed; this speed is, within the limits of the observational errors, the speed of light.[15]

---

[14]Riemann (1867, p. 243):

Nach der Bestimmung von Weber und Kohlrausch ist

$$c = 439\,450 \times 10^5 \frac{\text{Millimeter}}{\text{Secunde}}$$

woraus sich α zu 41 949 geographischen Meilen in der Secunde ergiebt während für die Lichtgeschwindigkeit von Busch aus Bradley's Aberrationsbeobachtungen 41 994 Meilen, und von Fizeau durch directe Messung 41 882 Meilen gefunden worden sind.

[15]Riemann (1867, p. 237):

Ich habe gefunden, dass die elektrodynamischen Wirhungen galvanischer Ströme sich erklären lassen, wenn man annimmt, dass die Wirkung einer elektrischen Masse auf die übrigen nicht momentan geschieht, sondern sich mit einer constanten (der Lichtgeschwindigkeit innerhalb der Gränzen der Beobachtungsfehler gleichen) Geschwindigkeit zu ihnen fortpflanzt.

Unfortunately, according to a remark by Clausius,[16] the analysis of B. Riemann was certainly inaccurate. The editor of the works of Riemann, H. Weber, guesses, with all likelihood, that Riemann recognized the error, which prevented him from delivering his note to print.

In 1868, while the work of Riemann was still unknown, the University of Bonn was celebrating its fiftieth anniversary jubilee. As *Gratulationsschrift* of the University of Tübingen, Carl Neumann presented a paper entitled. *Theoria nova phænomenis electricis applicanda*; this writing contained a summary of a theory that was [193] later published *in extenso* under this title[17]: *Die Principien der Elektrodynamik*.

The fundamental hypothesis of Carl Neumann essentially matched that of Riemann; the author stated in these words:

> A new supposition is introduced in making this motive cause, which we call potential, not be immediately but gradually transmitted in time from one mass to another, and—like light—to propagate with a great and constant speed. We will denote this speed by the letter c.
>
> This supposition, together with the other, the supreme and sacrosanct principle principle of Hamilton meeting no exceptions, is made *fundamental to our theory*, from which (without any further supposition) those well-founded laws of the celebrated AMPÈRE, NEUMANN, and WEBER, on their own foundations, will spontaneously emanate.[18]

But if the essential hypothesis accepted by Carl Neumann agrees with what B. Riemann issued, it immediately opposes it when its author translates it into formulas.

Consider, he says, two points M, M' carrying electric charges and acting on each other. Let r be their distance at time t. From what we have said about the propogation of the potential, we must distinguish two species of potential: the *emissive potential* and the *receptive potential*.

The *emissive potential* of the point M is the potential that the point M emits at time t, and which only reaches the point M' some time later. Its expression is

$$\omega_0 = \frac{ee'}{r}.$$

---

[16]R. Clausius, POGGENDORFF'S ANNALEN, Bd. GXXXV, p. 606; 1869.

[17]C. Neumann, *Die Principien der Elektrodynamik*, MATHEMATISCHE ANNALEN, Bd. XVII, p. 400.

[18]Neumann (1868, p. 121)

Nova introducitur suppositio, statuendo, causam illam motricem, quam potentiale vocamus, ab altéra massa ad alteram non subito sed progiediente tempore transmitti, atque—ad instar lucis—per spatium propagari celeritate quadam permagna et constante. Quam celeritatem denotabimus litera c.

Ista suppositio, conjuncta cum hac altera, principium Hamiltonianum normam exprimere supremam ac sacrosanctam nullis exceptionibus obviam, fit *suppositio in theoria nostra fundamentalis*, ex qua absque ulla ulteriore suppositione leges illæ notissimæ a cel[is], AMPÈRE, NEUMANN, WEBER, conditæ sua sponte emanabunt.

Carl Neumann defines the *receptive potential* as follows:

We will call the *receptive potential* that which any point *receives at time t* that had sometime before been emitted by another point. Wherefore, it is clear that the receptive potential formed with respect to any *given* time [194] is also the same as the emissive potential formed with respect to any *prior* time.[19]

By considerations that would take too long to explain here, but can be found in the paper entitled: *Die Principien der Elektrodynamik*, Carl Neumann arrives at the expression of receptive potential ω which gives the following equalities:

$$\omega = w + \frac{d\pi}{dt},$$

$$w = \frac{ee'}{r}\left[1 + \frac{1}{c^2}\left(\frac{dr}{dt}\right)^2\right],$$

$$\pi = ee'\left[\frac{\log r}{c} - \frac{1}{2c^2}\left(\frac{dr}{dt}\right)^2\right].$$

From this expression of the emissive potential, Hamilton's principle allows one to derive the expression of the force that each point experiences at time t. This force is directed along the line that joins two points, is repulsive, and has the magnitude

$$\frac{ee'}{r^2}\left[1 - \frac{1}{c^2}\left(\frac{dr}{dt}\right)^2 + \frac{2}{c^2}r\frac{d^2r}{dt^2}\right].$$

This is the force given by Weber's law.

So that Carl Neumann's theory is consistent with the known laws of electrodynamics, it will be necessary to give the constant c the value, determined by Weber and Kohlrausch,

$$c = 439\,450 \times 10^6 \frac{\text{millimeters}}{\text{second}}.$$

So, the potential does not propagate with a speed equal to the speed of light V in a vacuum, but with a larger speed equal to $V\sqrt{2}$.

---

[19]Neumann (1868, p. 121):

Potentiale receptivum vocabimus id, quod utrumque punctum *recipit tempore t*, aliquanto antea ab altero puncto emissum. Unde elucet potentiale receptivum respectu *dati* temporis cujuslibet formatum idem esse ac potentiale emissivum respectu temporis cujusdam *prioris* formatum.

In the same volume of the *Poggendorff's Annalen* where [195] the electrodynamics of B. Riemann was printed for the first time, L. Lorenz published[20] a theory that had, with the thought of Riemann, and unbeknown to the author, a closer affinity with the theory of C. Neumann.

In generalizing by induction the equations of the electrodynamics given by W. Weber, G. Kirchhoff[21] came to a system of equations governing the propagation of electric actions in conductive bodies,

Let $V = \sum \frac{q}{r}$ be the electrostatic potential function, where the summation extends to all charges electrical charges q of the system.

This function can be expressed more explicitly.

At time t, at the point (x, y, z) of an electrified volume, the solid electrical density is $\sigma(x, y, z, t)$; at time t, at the point (x, y, z) of an electrified surface, the electric surface density is $\Sigma(x, y, z, t)$. We then have

$$V(x, y, z, t) = \int \frac{\sigma(x', y', z', t)}{r} d\omega' + \int \frac{\Sigma(x', y', z', t)}{r} dS', \qquad (8.3a)$$

the first integral extending over all the elements $d\omega'$ of the electrified volumes and the second integral extending over all elements dS' of the electrified surfaces.

Let

$$u(x, y, z, t), \quad v(x, y, z, t), \quad w(x, y, z, t)$$

be the three components of the electric current[22] at the point (x, y, z), at time t. [196]

Consider the functions

$$\begin{cases} U(x, y, z, t) = \int \frac{x' - x}{r^3}[(x' - x)u(x', y', z', t) \\ \qquad\qquad\qquad + (y' - y)v(x', y', z', t) \qquad\qquad (8.4) \\ \qquad\qquad\qquad + (z' - z)w(x', y', z', t)]d\omega', \\ V(x, y, z, t) = \dots, \quad W(x, y, z, t) = \dots . \end{cases}$$

---

[20]L. Lorenz, *Sur l'identité des vibrations de la lumière et des courants électriques* (cf. SELSKABS. OVERS., 1867, p, 26.—POGGENDORFF'S ANNALEN, Bd. CXXXI, p. 243; 1867.—ŒUVRES SCIENTIFIQUES DE L. LORENZ, revised and annotated by H. Valentinier, t. 1, p. 173; 1896). [English translation: Lorenz (1867)].

[21]G. Kirchhoff, *Ueber die Bewegung der Elektricität in Leitern.* (POGGENDORFF'S ANNALEN, Bd. CII, 1857). [English translation: Graneau and Assis (1994)].

[22]In the memoir of G. Kirchhoff, u, v, w, have slightly different meanings, linked to particular conceptions of Weber on the nature of power.

The equations of motion of electricity in a conductive body, of which ρ is the specific resistance, are written, according to G. Kirchhoff,

$$\begin{cases} u = -\dfrac{1}{\rho}\left(\dfrac{\partial V}{\partial x} + \dfrac{2}{c^2}\dfrac{\partial U}{\partial t}\right), \\[2mm] v = -\dfrac{1}{\rho}\left(\dfrac{\partial V}{\partial y} + \dfrac{2}{c^2}\dfrac{\partial v}{\partial t}\right), \\[2mm] w = -\dfrac{1}{\rho}\left(\dfrac{\partial V}{\partial z} + \dfrac{2}{c^2}\dfrac{\partial w}{\partial t}\right). \end{cases} \tag{8.5}$$

L. Lorenz rightly notes that in taking as point of departure not not the formulas of induction that Weber gives, but other formulas that are strictly equivalent to them in the only case hitherto studied, those of uniform linear currents, one cannot obtain the preceding equations, but some other analogous equations, in particular these:

$$\begin{cases} u = -\dfrac{1}{\rho}\left(\dfrac{\partial V}{\partial x} + \dfrac{2}{c^2}\dfrac{\partial F}{\partial t}\right), \\[2mm] v = -\dfrac{1}{\rho}\left(\dfrac{\partial V}{\partial y} + \dfrac{2}{c^2}\dfrac{\partial G}{\partial t}\right), \\[2mm] w = -\dfrac{1}{\rho}\left(\dfrac{\partial V}{\partial z} + \dfrac{2}{c^2}\dfrac{\partial H}{\partial t}\right), \end{cases} \tag{8.6a}$$

[197] where we have

$$\begin{cases} F(x, y, z, t) = \displaystyle\int \dfrac{u(x', y', z', t)}{r}\,d\omega', \\[3mm] G(x, y, z, t) = \displaystyle\int \dfrac{v(x', y', z', t)}{r}\,d\omega', \\[3mm] H(x, y, z, t) = \displaystyle\int \dfrac{w(x', y', z', t)}{r}\,d\omega'. \end{cases} \tag{8.7a}$$

This remark was soon to be picked up by Helmholtz[23] and suggests to him to introduce into the electrodynamic theories the numeric constant of such great importance, which he designates by the letter k.

---

[23]Helmholtz, *Ueber die Gesetze der inconstanten elektrischen Ströme in körperlich ausgedehnten Leitern* (VERHANDLUNGEN DES NATURHISTORISCH-MEDICINISCHEN VEREINS ZU HEIDELBERG, 21 January 1870.—WISSENSCHAFTLICHE ABHANDLUNGEN, Bd. I, p. 537).—*Ueber die Bewegungsgleichungen der Elektrodynamik für ruhende leitende Körper* (BORCHARDT'S JOURNAL FÜR REINE UND ANGEWANDTE MATHEMATIK, Bd. LXXII, p. 57.—WISSENSCHAFTLICHE ABHANDLUNGEN, Bd. I, p. 545).

These are Eq. (8.6a) that L. Lorenz takes as equations of motion of the electricity; but instead of keeping the functions V, F, G, H defined by the equalities (8.3a) and (8.7a), he substitutes for them the functions

$$\overline{V}(x, y, z, t) = \int \frac{\sigma(x', y', z', t - \frac{r}{a})}{r} d\omega'$$
$$+ \int \frac{\Sigma(x', y', z', t - \frac{r}{a})}{r} dS', \tag{8.3b}$$

$$\begin{cases} \overline{F}(x, y, z, t) = \int \frac{u(x', y', z', t - \frac{r}{a})}{r} d\omega', \\[2mm] \overline{G}(x, y, z, t) = \int \frac{v(x', y', z', t - \frac{r}{a})}{r} d\omega', \\[2mm] \overline{H}(x, y, z, t) = \int \frac{w(x', y', z', t - \frac{r}{a})}{r} d\omega', \end{cases} \tag{8.7b}$$

[198] where

$$a = \frac{c}{\sqrt{2}}. \tag{8.8}$$

It is, we see, the hypothesis issued by B. Riemann, whereby the electric potential function propagates with the speed a, that L. Lorenz admits, and that he extends to the functions F, G, H, components of the *electrotonic state*.

Equation (8.6a) become

$$\begin{cases} u = \frac{1}{\rho} \left(\frac{\partial \overline{U}}{\partial x} + \frac{2}{c^2} \frac{\partial \overline{F}}{\partial t} \right), \\[3mm] v = \frac{1}{\rho} \left(\frac{\partial \overline{V}}{\partial x} + \frac{2}{c^2} \frac{\partial \overline{G}}{\partial t} \right), \\[3mm] w = \frac{1}{\rho} \left(\frac{\partial \overline{W}}{\partial x} + \frac{2}{c^2} \frac{\partial \overline{H}}{\partial t} \right). \end{cases} \tag{8.6b}$$

These equations only differ from Eq. (8.6a) by the substitution of $(t - \frac{r}{a})$ for t. Now, in all experiments r is equal to more than a few meters, while a represents a velocity roughly equal to $300\,000$ km/s; $(t - \frac{r}{a})$ thus differs extremely little from t and Eqs. (8.6a) and (8.6b) can be viewed as also verified by experience.

It is easily checked that at any point of a continuous mass, we have

$$a^2 \Delta \overline{V} - \frac{\partial^2 \overline{V}}{\partial t^2} = -4\pi a^2 \sigma(x, y, z, t),$$

$$a^2 \Delta \overline{F} - \frac{\partial^2 \overline{F}}{\partial t^2} = -4\pi a^2 u(x, y, z, t),$$

$$a^2 \Delta \overline{G} - \frac{\partial^2 \overline{G}}{\partial t^2} = -4\pi a^2 v(x, y, z, t),$$

$$a^2 \Delta \overline{H} - \frac{\partial^2 \overline{H}}{\partial t^2} = -4\pi a^2 w(x, y, z, t).$$

[199] Therefore, it is not difficult to see that Eqs. (8.6a) and (8.8) allow us to write the equations

$$
\begin{cases}
\Delta u - \dfrac{2}{c^2} \dfrac{\partial^2 u}{\partial t^2} = \dfrac{4\pi}{\rho} \left(\dfrac{\partial \sigma}{\partial x} + \dfrac{2}{c^2} \dfrac{\partial u}{\partial t} \right), \\[3mm]
\Delta v - \dfrac{2}{c^2} \dfrac{\partial^2 v}{\partial t^2} = \dfrac{4\pi}{\rho} \left(\dfrac{\partial \sigma}{\partial y} + \dfrac{2}{c^2} \dfrac{\partial v}{\partial t} \right), \\[3mm]
\Delta w - \dfrac{2}{c^2} \dfrac{\partial^2 w}{\partial t^2} = \dfrac{4\pi}{\rho} \left(\dfrac{\partial \sigma}{\partial z} + \dfrac{2}{c^2} \dfrac{\partial w}{\partial t} \right),
\end{cases}
\tag{8.9}
$$

to which must be joined the continuity equation

$$\frac{\partial u}{\partial x} + \frac{\partial v}{\partial y} + \frac{\partial w}{\partial z} + \frac{\partial \sigma}{\partial t} = 0.$$

One easily sees that each of the three quantities

$$\omega_x = \frac{w}{y} - \frac{\partial v}{\partial z}, \qquad \omega_y = \frac{u}{z} - \frac{\partial w}{\partial x}, \qquad \omega_z = \frac{v}{x} - \frac{\partial u}{\partial y}$$

satisfies the equation

$$\Delta \omega - \frac{2}{c^2} \frac{\partial^2 \omega}{\partial t^2} = \frac{8\pi}{\rho c^2} \frac{\partial \omega}{\partial t}.$$

If the medium under consideration is extremely resistant, so that ρ has a very high value, the second member of this equation is negligible compared to the first member; the equation reduces to the well-known form

$$\Delta \omega - \frac{2}{c^2} \frac{\partial^2 \omega}{\partial t^2} = 0,$$

which teaches us that, in the medium considered, the transverse electric current propagates with speed $\frac{c}{\sqrt{2}}$. We thus arrive at the following proposition:

In an extremely resistant medium, the transverse electric current propagates with a speed equal to the speed of light in a vacuum. [200]

Encouraged by this important result, L. Lorenz did not hesitate to formulate an electromagnetic theory of light: all transparent media are very poor conductors of electricity, and the light that propagates in these media consists of periodic transverse electric currents.

The hypothesis is certainly seductive; it, however, faces great difficulties.

In the first place, the equations obtained do not exclude the possibility of longitudinal electric currents, whose role will be difficult to explain.

Secondly, and this is the most serious objection: according to the previous theory, in any very poor conductive medium, transverse electric currents always propagate with a speed equal to the speed of light in a vacuum. On the contrary, in a transparent medium, light travels with a speed that characterizes this medium and which is less than the speed of light in a vacuum; and we see no easy way to change the hypothesis of the previous theory so that this contradiction disappears.

This contradiction seems to condemn irrevocably the electromagnetic theory of light proposed by L. Lorenz.

8.3 The Fundamental Hypothesis of Maxwell—Electrodynamic Polarization of Dielectrics

An extremely deep logical difference separates the hypotheses of B. Riemann, L. Lorenz, and C. Neumann from the hypotheses on the propagation of physical actions thus far admitted.

The theory of the emission of light represented the propagation of light as like the trajectory of a projectile; what propagated, in this theory, was a substance.

The propagation of sound occurs, on the contrary, without the substance serving this propagation, air for example, undergoing significant displacements; but, while a mass of air, initially moving, falls back to rest, a nearby mass, which was at rest, is put in motion. In this case, there is propagation, not of a substance, but of an accident,[24] of a movement. [201]

These two types are similar to most physical theories involving the notion of propagation. In the theory of waves, the transmission of light is the propagation of a movement; and when adopting the ideas of Weber, Kirchhoff studies the propagation of electricity in conductive bodies, he considers it as the flow of a certain substance.

We can obviously generalize further and conceive the propagation in a body of an accident which would not be a movement of this body, but of some quality. For a physicist who regards electricity neither as a fluid nor a movement, but simply a certain quality, Kirchhoff's equations represent a propagation of this quality through conductive bodies.

But all these different ways of considering the concept of propagation have a common character; substance or accident, it is something real that disappears in a region of space in order to appear in a nearby area. It is not the case in the theories of the propagation of electric actions proposed by B. Riemann, L. Lorenz, and Carl Neumann; there is not a reality that travels through the space, but a fiction, a mathematical symbol, such as the potential function or the components of the electrotonic state.

---

[24][A property in the Aristotelian sense].

This character of the new theories, perhaps, was suspected by L. Lorenz. In any case, it was clearly perceived by Carl Neumann; but he does not hesitate to consider the potential function, whose propagation he assumes, as a reality. He said:

> It is well known that for given forces there is a given potential, and vice versa, for a given potential, there are given forces. Wherefore, nothing new appears in traditional mechanical theory for causing, the potential being the principal cause, it to produce the forces; viz., the potential is called the *real cause of the motion*, but forces only express the form or species by the cause which produced them.[25]

This passage would allow us, I think, quite rightly to regard Carl Neumann as the creator of the philosophical and scientific doctrine that is now in such great vogue as the doctrine of the migration of energy (*Wanderung der Energie*).

The ideas of Maxwell have nothing in common with these doctrines; [202] mathematical symbols do not propagate. For example, the expression of the instantaneous electrostatic potential function at the point (x, y, z) in a medium of dielectric strength K is

$$V(x, y, z, t) = \frac{1}{K} \sum \frac{q(x', y', z', t)}{r}$$

and not, as the hypothesis of B. Riemann would have it,

$$V(x, y, z, t) = \frac{1}{K} \sum \frac{1}{r} q\left(x', y', z', t - \frac{r}{a}\right).$$

What is propagating is a real quality: in conductive bodies, the conduction current; in dielectric bodies, the displacement flux.

The consideration of dielectric bodies is, moreover, one of the essentially new points of Maxwell's theory. B. Riemann, nor C. Neumann, made the slightest allusion to the polarization of dielectrics; for L. Lorenz, insulating bodies are simply bodies whose specific resistance is very large, poorly conducting bodies,[26] and it is to the *conduction current* propagating in similar bodies that he likens the light vibration.

On the contrary, for Maxwell, light that propagates in transparent bodies consists essentially in *displacement currents* produced inside the dielectric body.

---

[25]Neumann (1868, p. 121)

> Potentiis datis datum esse potentiale, ac vice versa, potentiali dato, datas esse potentias, satis notum est. Unde apparet in traditam mechanices theoriam nil novi introduci statuendo, potentiale principalem esse causam, ab isto procreari potentias, scilicet potentiale vocare *veram causam motricem*, potentias vero tantummodo formam vel speciem exprimere ab illa causa sibi paratam.

[26]The difference between the point of view of Maxwell and the point of view of Lorenz was very well marked in a note added by H. Valentinier to the scientific works of the latter (L. Lorenz, ŒUVRES SCIENTIFIQUES, revised and annotated by H. Valentiner, tome I, p. 204, note 16).

These displacement currents, we know, produce the same ponderomotive and electromotive actions as the the conduction current; but their generation is subject to another law, and the invention of this law is one of the most powerful and most productive of Maxwell's ideas.

In a system where equilibrium is established, the components f, g, h [203] of the displacement are related to the derivatives of the electrostatic potential function ψ by the equalities [1st Part, equalities (5.15)]

$$f = -\frac{K}{4\pi}\frac{\psi}{x}, \quad g = -\frac{K}{4\pi}\frac{\psi}{y}, \quad h = -\frac{K}{4\pi}\frac{\psi}{z}.$$

In a system that is not in equilibrium, the previous equalities should be replaced by

$$f = \frac{K}{4\pi}E_x, \quad g = \frac{K}{4\pi}E_y, \quad h = \frac{K}{4\pi}E_z, \tag{8.10}$$

where E_x, E_y, E_z are the components of the total electromotive field, as well as the induction field and the static field.

We see this idea arises naturally from the hypotheses admitted by Maxwell regarding the constitution of dielectrics.

We have recognized, in the course of this study, that Maxwell was almost constantly guided, in his suppositions regarding dielectrics, by the hypotheses of Faraday and Mossotti, themselves designed in imitation of the magnetic hypotheses of Poisson. According to these hypotheses, a dielectric is formed of small conductive masses embedded in an insulating cement. The action of an electromotive induction field on a dielectric field will therefore result in actions that this field exerts on a large number of open[27] conductors.

However, in an open conductor, an electromotive induction field produces the same effect as a static electromotive field; it requires electricity to be distributed so that the positive charge builds up on one of the ends of the conductor and the negative charge at the other end; in other words, this field polarizes the open conductor.

Maxwell insists repeatedly about this action that an induction field exerts on an open conductor.

He already wrote in his memoir *On Faraday's Lines of Force*[28]:

> Let us take as another example the case of a linear conductor, not forming a closed circuit, and let it be made to traverse the lines of magnetic [204] force, either by its own motion, or by changes in the magnetic field. An electromotive force will act in the direction of the conductor, and, as it cannot produce a current, because there is no circuit, it will produce electric tension at the extremities.

From this passage, Maxwell makes, for the moment, no conclusion related to the polarization of dielectrics, to which he is hardly attached in this first memoir on electricity; it is otherwise in the memoir: *On Physical Lines of Force*.

---

[27] [In the sense of "open circuit"].

[28] J. Clerk Maxwell, Scientific Papers, vol. I, p. 186.

He wrote[29]:

We know by experiment that electric tension is the same thing, whether observed in statical or in current electricity; so that an electromotive force produced by magnetism may be made to charge a Leyden jar, as is done by the coil machine.

...When a difference of tension exists in different parts of any body, the electricity passes, or tends to pass, from places of greater to places of smaller tension.

The application of these considerations to small conductive bodies that contain a dielectric is immediate; it imposes conclusions that Maxwell states in these terms[30]:

Electromotive force acting on a dielectric produces a state of polarization of its parts similar in distribution to the polarity of the particles of iron under the influence of a magnet,[31] and, like the magnetic polarization, capable of being described as a state in which every particle has its poles in opposite conditions.

In a dielectric under induction, we may conceive that the electricity in each molecule is so displaced that one side is rendered positively, and the other negatively electrical, but that the electricity remains entirely connected with the molecule, and does not pass from one molecule to another.

The effect of this action on the whole dielectric mass is to produce a general displacement of the electricity in a certain direction...The amount of the displacement depends on the nature of the [205] body, and on the electromotive force; so that if h is the displacement, R the electromotive force, and E a coefficient depending on the nature of the dielectric,

$$R = -4\pi E^2 h;$$

and if r is the value of the electric current due to displacement,[32]

$$r = \frac{dh}{dt}.$$

The same ideas are found, in a still sharper form, in the memoir: *A Dynamical Theory of the Electromagnetic Field*.

Maxwell writes[33]:

When a body is moved across the lines of magnetic force it experiences what is called an electromotive force; the two extremities of the body tend to become oppositely electrified, and an electric current tends to flow through the body. When the electromotive force is sufficiently powerful, and is made to act on certain compound bodies, it decomposes them, and causes one of their components to pass towards one extremity of the body, and the other in the opposite direction. [206]

Here we have evidence of a force causing an electric current in spite of resistance; electrifying the extremities of a body in opposite ways, a condition which is sustained only by the action of the electromotive force, and which, as soon as that force is removed, tends, with an equal and opposite force, to produce a counter current through the body and to restore the

---

[29] J. Clerk Maxwell, SCIENTIFIC PAPERS, vol. I, p. 490.

[30] J. Clerk Maxwell, *loc. cit.*, p. 491.

[31] [Maxwell's footnote:] See Prof. Mossotti, "Discussione Analitica," *Mem. della Soc. Italiana* (Modena), Vol. XXIV. Part 2, p. 49.

[32] Regarding the sign of the second member, see 1st Part, Eq. (4.1).

[33] J. Clerk Maxwell, SCIENTIFIC PAPERS, vol. I, p. 530.

original electrical state of the body; and finally, if strong enough, tearing to pieces chemical compounds and carrying their components in opposite directions, while their natural tendency is to combine, and to combine with a force which can generate an electromotive force in the reverse direction.

This, then, is a force acting on a body caused by its motion through the electromagnetic field, or by changes occurring in that field itself; and the effect of the force is either to produce a current and heat the body, or to decompose the body, or, when it can do neither, to put the body in a state of electric polarization, a state of constraint in which opposite extremities are oppositely electrified, and from which the body tends to relieve itself as soon as the disturbing force is removed.

...When electromotive force acts on a conducting circuit, it produces a current...But when electromotive force acts on a dielectric it produces a state of polarization of its parts...

and Maxwell, also citing Faraday and Mossotti,[34] who visibly inspired him, reproduced, on the subject of this dielectric polarization, the passage of the memoir: *On Physical Lines of Force* that we have quoted above.

These are, in their natural sequence, the inductions which led Maxwell to pose the general equations of the dielectric polarization[35]

$$ f = \frac{K}{4\pi} E_x, \quad g = \frac{K}{4\pi} E_y, \quad h = \frac{K}{4\pi} E_z. $$

In a homogeneous medium, the components E_x, E_y, E_z of the electromotive field are given by equalities (7.56), such that equalities (8.10) become

$$ \begin{cases} f = -\dfrac{K}{4\pi} \left(\dfrac{\partial \Psi}{\partial x} + \dfrac{\partial F}{\partial t} \right), \\[2mm] g = -\dfrac{K}{4\pi} \left(\dfrac{\partial \Psi}{\partial y} + \dfrac{\partial G}{\partial t} \right), \\[2mm] h = -\dfrac{K}{4\pi} \left(\dfrac{\partial \Psi}{\partial z} + \dfrac{\partial H}{\partial t} \right). \end{cases} \tag{8.11} $$

[207] Moreover, in this case, the components of the displacement current have the values

$$ \overline{u} = \frac{\partial A}{\partial t}, \quad \overline{v} = \frac{\partial B}{\partial t}, \quad \overline{w} = \frac{\partial C}{\partial t}. \tag{6.3} $$

---

[34] [Faraday, *Experimental Researches*, Series XI.; Mossotti, *Mem. della Soc. Italiana* (Modena), Vol. XXIV. Part 2, p. 49.].

[35] J. Clerk Maxwell, *A Dynamical Theory of the Electromagnetic Field*, (SCIENTIFIC PAPERS, vol. I, p. 560.)—*Treatise on Electricity and Magnetism*, vol. II, p. 287 [232].

We there have

$$\begin{cases} \overline{u} = -\dfrac{K}{4\pi}\dfrac{\partial}{\partial t}\left(\dfrac{\partial \Psi}{\partial x} + \dfrac{\partial F}{\partial t}\right), \\[2mm] \overline{v} = -\dfrac{K}{4\pi}\dfrac{\partial}{\partial t}\left(\dfrac{\partial \Psi}{\partial y} + \dfrac{\partial G}{\partial t}\right), \\[2mm] \overline{w} = -\dfrac{K}{4\pi}\dfrac{\partial}{\partial t}\left(\dfrac{\partial \Psi}{\partial z} + \dfrac{\partial H}{\partial t}\right). \end{cases} \tag{8.12}$$

These equations are the basis of the electromagnetic theory of light.

8.4 First Draft of the Electromagnetic Theory of Light of Maxwell

However, before developing an electromagnetic theory of light based on these equations, Maxwell obtained two essential laws of this theory by a completely different method. This method, closely linked to the mechanical hypotheses contained in the memoir: *On Physical Lines of Force*, is set out in this memoir.

We have seen (1st Part, Chap. 4) how, in this memoir, Maxwell represents the action of an electromotive field in a dielectric. The electromotive force is considered as a push that is exerted on the perfectly elastic cell walls. If R is the electromotive field, the walls undergo a displacement in the direction of this field; the average value per unit of volume of this movement, that he designates with the letter h, is linked to the electromotive field R by the relationship [1st Part, equality (4.1b)]

$$R = 4\pi E^2 h,$$

[208] E^2 being a quantity that depends on the elasticity of the cell walls.

Without discussing, from the point of view of the theory of elasticity, the solution of the problem addressed by Maxwell, we confine ourselves to indicate the relationship that exists, according to him, between E^2 and the coefficients of elasticity of the substance.

Maxwell expresses[36] E^2 in terms of two coefficients that he designates by μ and m and that, to avoid confusion, we will refer to by μ' and m; this expression is as follows:

$$E^2 = \pi m \frac{9\mu'}{3\mu' + 5m}. \tag{8.13}$$

The coefficient μ' is defined[37] as the ratio of pressure to the cube of the contraction in a uniformly pressed body; it is thus the reverse of what is usually called the cube

---

[36] J. Clerk Maxwell, SCIENTIFIC PAPERS, vol. I, p. 495, equality (107).

[37] J. Clerk Maxwell, *loc. cit.*, p. 493, equality (80).—To make this equality agree with the rest of the presentation by Maxwell, he must change the sign of the second member.

of the coefficient of compressibility. If we designate by Λ and M the coefficients that Lamé designates by λ and μ, we will have[38]

$$\mu' = \frac{3\Lambda + 2M}{3}. \tag{8.14}$$

As for the coefficient m, in comparing[39] Maxwell's equations to those of Lamé, we find

$$m = 2M. \tag{8.15}$$

In virtue of equalities (8.14) and (8.15), equality (8.13) becomes

$$E^2 = \pi m \frac{3\Lambda + 2M}{\Lambda + 4M}. \tag{8.16}$$

[209] If one accepts the theory of molecular elasticity as Poisson has developed it, we have, as is known, the equality

$$\Lambda = M \tag{8.17}$$

and equality (8.16) becomes

$$E^2 = \pi m, \tag{8.18}$$

which Maxwell accepts[40] for the further development of his theory.

According to this theory, two electrical charges whose values in electromagnetic units are q_1, q_2 are repelled at a distance r with a force [1st Part, equality (4.37)]

$$F = E^2 \frac{q_1 q_2}{r^2}, \tag{8.19}$$

E^2 having the appropriate value for the interposed dielectric.

If the dielectric is the vacuum, the value of E^2 can be obtained from the famous experience of Weber and Kohlrausch. We then find[41] that E is a quantity of same species as a speed whose numerical value is

$$E = 310\,740 \times 10^6 \frac{\text{millimeters}}{\text{second}}. \tag{8.20}$$

---

[38]Lamé, *Leçons sur l'élasticité*, 2nd edition, p. 74, equality (a).

[39]J. Clerk Maxwell, *loc. cit.*, p. 493, equality (83) and Lamé, *loc. cit.*, p. 65, equalities (1).

[40]J. Clerk Maxwell, *loc. cit.*, p. 495, equality (108).

[41]J. Clerk Maxwell, *loc. cit.*, p. 499, equality (131).

Reaching this point, Maxwell continues[42] in these terms:

> To find the rate of propagation of transverse vibrations through the elastic medium of which the cells are composed, on the supposition that its elasticity is due entirely to forces acting between pairs of particles.[43] [210]
>
> By the ordinary method of investigation we know that

$$V = \sqrt{\frac{m}{\rho}}, \tag{8.21}$$

> where m is the coefficient of transverse elasticity, and ρ is the density.

The density that must be included in this formula is the density of the elastic medium which forms the walls of the cells. Without telling us about this transposition, Maxwell assumes that ρ refers to the density of the fluid that fills the cells and then admits the relationship

$$\mu = \pi \rho \tag{8.22}$$

that he was led to establish[44] between this density and the magnetic permeability μ. He then found

$$\mu V^2 = \pi m$$

or, in virtue of equality (8.18),

$$E = V \sqrt{\mu}. \tag{8.23}$$

He comments[45] on this result in these terms:

In air or vacuum $\mu = 1$ and therefore

$$V = E$$
$$= 310\,740 \times 10^6 \text{ millimeters per second}$$
$$= 193\,088 \text{ miles per second.}$$

The velocity of light in air, as determined by M. Fizeau,[46] is 70 843 leagues per second (25 leagues to a degree) which gives

$$V = 314\,858 \times 10^6 \text{ millimeters per second}$$
$$= 195\,647 \text{ miles per second.}$$

---

[42] J. Clerk Maxwell, loc. cit., p. 499.

[43] By these words, Maxwell refers to the molecular theory of Poisson.

[44] J. Clerk Maxwell, loc. cit., pp. 456 and 457.

[45] J. Clerk Maxwell, loc. cit., pp. 499 and 500.

[46] [Maxwell's footnote:] Comptes Rendus, Vol. MIX. (1849), p. 90. In Galbraith and Haughton's Manual of Astronomy, M. Fizeau's result is stated at 169 944 geographical miles of 1000 fathoms, which gives 193 118 statute miles; the value deduced from aberration is 192 000 miles.

The velocity of transverse undulations in our hypothetical medium, calculated from the electro-magnetic experiments of MM. Kohlrausch and Weber, agrees so exactly with the velocity of light calculated from the optical experiments of M. Fizeau, that we can scarcely avoid the inference that *light consists in the transverse undulations of the same medium which is the cause of electric and magnetic phenomena.*

The capacitance of a plane capacitor of surface S, whose armatures are separated by a thickness θ of a given dielectric 1, has the value [1st part, equality (4.46)]

$$C_1 = \frac{1}{4\pi E_1^2} \frac{S}{\theta}.$$

If the space between the two plates of the capacitor is empty, this capacitor has, similarly, capacitance

$$C = \frac{1}{4\pi E^2} \frac{S}{\theta}.$$

The ratio

$$D_1 = \frac{C_1}{C}$$

is, by definition, the specific inductive capacity of dielectric 1. So we have

$$D_1 = \frac{E^2}{E_1^2}$$

or, in virtue of equality (8.23),

$$D_1 = \frac{V^2}{V_1^2} \frac{1}{\mu_1}. \tag{8.24}$$

...so[47] that the inductive capacity of a dielectric varies directly as the square of the index of refraction, and inversely as the magnetic inductive capacity.

Thus, from 1862 until the note by Bernhard Riemann was published, while the theories of L. Lorenz and C. Neumann were not yet conceived, Maxwell was already in possession of the essential laws of the electromagnetic theory of light. Unfortunately, the method by which it was reached, very [212] different from the one he has since proposed, was tainted by a serious clerical error. In virtue of equality (8.15), equality (8.21) would become

$$V = \sqrt{\frac{2M}{\rho}},$$

---

[47] J. Clerk Maxwell, *loc. cit.*, p. 501.

an incorrect formula for which one must substitute the equality[48]

$$V = \sqrt{\frac{M}{\rho}}.$$

8.5 Final Form of the Electromagnetic Theory of Light of Maxwell

Twice Maxwell has explained, with variations in detail, the electromagnetic theory of light in an accurate form free from mechanical hypotheses: first,[49] in the memory: *A Dynamical Theory of the Electromagnetic Field*; a second time,[50] in the *Treatise on Electricity and Magnetism*.

Consider the system of six equations of Maxwell

$$\begin{cases} \dfrac{\partial \gamma}{\partial y} - \dfrac{\partial \beta}{\partial z} = -4\pi (u + \overline{u}), \\[2mm] \dfrac{\partial \alpha}{\partial z} - \dfrac{\partial \gamma}{\partial x} = -4\pi (v + \overline{v}), \\[2mm] \dfrac{\partial \beta}{\partial x} - \dfrac{\partial \alpha}{\partial y} = -4\pi (w + \overline{w}). \end{cases} \tag{7.5}$$

$$\begin{cases} \dfrac{\partial H}{\partial y} - \dfrac{\partial G}{\partial z} = -\mu\alpha, \\[2mm] \dfrac{\partial F}{\partial z} - \dfrac{\partial H}{\partial x} = -\mu\beta, \\[2mm] \dfrac{\partial G}{\partial x} - \dfrac{\partial F}{\partial y} = -\mu\gamma. \end{cases} \tag{7.54b}$$

We put (7.55b) [213]

$$\frac{\partial F}{\partial x} + \frac{\partial G}{\partial y} + \frac{\partial H}{\partial z} = J, \tag{7.55b}$$

---

[48] Lamé, *loc. cit.*, p. 142, equality (9).

[49] J. Clerk Maxwell, SCIENTIFIC PAPERS, vol. I, pp. 577–588.

[50] J. Clerk Maxwell, *Treatise on Electricity and Magnetism*, trad. française, t. II, pp. 485–504 [383–398].

supposing the medium is homogeneous. We will easily obtain three equalities

$$
\begin{cases}
\Delta F = \dfrac{\partial J}{\partial x} - 4\pi\mu(u + \bar{u}), \\[2mm]
\Delta G = \dfrac{\partial J}{\partial y} - 4\pi\mu(v + \bar{v}), \\[2mm]
\Delta H = \dfrac{\partial J}{\partial z} - 4\pi\mu(w + \bar{w}).
\end{cases}
\tag{8.25}
$$

These equations are general. Now suppose the medium is not conductive or dielectric. We will have

$$
u = 0, \quad v = 0, \quad w = 0,
$$

while $\bar{w}$, $\bar{v}$, $\bar{w}$ will be given by equalities (8.12). Therefore, equalities (8.25) will become

$$
\begin{cases}
\Delta F - K\mu\dfrac{\partial^2 F}{\partial t^2} = \dfrac{\partial J}{\partial x} + K\mu\dfrac{\partial^2 \psi}{\partial x\,\partial t}, \\[2mm]
\Delta G - K\mu\dfrac{\partial^2 G}{\partial t^2} = \dfrac{\partial J}{\partial y} + K\mu\dfrac{\partial^2 \psi}{\partial y\,\partial t}, \\[2mm]
\Delta H - K\mu\dfrac{\partial^2 H}{\partial t^2} = \dfrac{\partial J}{\partial z} + K\mu\dfrac{\partial^2 \psi}{\partial z\partial t}.
\end{cases}
\tag{8.26}
$$

Together with equalities (7.54b), these relationships give us, in the first place, the equalities

$$
\begin{cases}
\Delta\alpha - K\mu\dfrac{\partial^2\alpha}{\partial t^2} = 0, \\[2mm]
\Delta\beta - K\mu\dfrac{\partial^2\beta}{\partial t^2} = 0, \\[2mm]
\Delta\gamma - K\mu\dfrac{\partial^2\gamma}{\partial t^2} = 0.
\end{cases}
\tag{8.27}
$$

[214] These three equations, whose form is well known, teach us that in a homogeneous dielectric, the three components α, β, γ of the magnetic field, which, according to equality (7.54b), satisfy the relationship

$$
\frac{\partial\alpha}{\partial x} + \frac{\partial\beta}{\partial y} + \frac{\partial\gamma}{\partial z} = 0
\tag{8.28}
$$

that characterizes the components of a transverse vibration propagating with a speed

$$
V = \sqrt{\frac{1}{K\mu}}.
\tag{8.29}
$$

The series of deductions of Maxwell is different in the memoir: *A Dynamical Theory of the Electromagnetic Field* and in the *Treatise on Electricity and Magnetism*. Let us adhere to the arguments expressed in the latter, which are more correct.

We differentiate the first equality (8.26) with respect to x, the second with respect to y, the third with respect to z and add, member by member, the obtained results by taking into account equality (7.55b); we find

$$K\mu\left(\frac{\partial}{\partial t}\Delta\Psi + \frac{\partial^2 J}{\partial t^2}\right) = 0. \tag{8.30}$$

On the other hand, equality (5.16) from the 1st Part teaches us that, in a homogeneous medium, the electric density e is given by the equality

$$K\Delta\Psi + 4\pi e = 0. \tag{8.31}$$

Finally, the equality

$$\frac{\partial u}{\partial x} + \frac{\partial v}{\partial y} + \frac{\partial w}{\partial z} + \frac{\partial e}{\partial t} = 0. \tag{6.19}$$

shows us that we have, in a non-conductive medium where

$$u = 0, \quad v = 0, \quad w = 0,$$

[215] the equality (8.32)

$$\frac{\partial e}{\partial t} = 0. \tag{8.32}$$

Equalities (8.30)–(8.32) give

$$\frac{\partial^2 J}{\partial t^2} = 0. \tag{8.33}$$

Hence[51] J must be a linear function of t, or a constant, or zero, and we may therefore leave J and Ψ out of account in considering periodic disturbances.

And Eq. (8.26) become, according to Maxwell,[52]

$$\begin{cases} \Delta F - K\mu\dfrac{\partial^2 F}{\partial t^2} = 0, \\[2mm] \Delta G - K\mu\dfrac{\partial^2 G}{\partial t^2} = 0, \\[2mm] \Delta H - K\mu\dfrac{\partial^2 H}{\partial t^2} = 0. \end{cases} \tag{8.34}$$

---

[51] J. Clerk Maxwell, *Treatise on Electricity and Magnetism*, t. II, p. 488 [385].

[52] J. Clerk Maxwell, *loc. cit.*, p. 488 [385], Eq. (9).

The sentence of Maxwell which we have cited, accurate with respect to the function J, is not for the function Ψ; but, without departing too much from the essential thought of Maxwell, one could reason as follows:

We differentiate equalities (8.26) twice with respect to t, taking into account equality (8.33) and the equality

$$\frac{\partial}{\partial t} \Delta \Psi = 0,$$

which results from equalities (8.31) and (8.32) and gives

$$\Delta \frac{\partial^2 \Psi}{\partial x \, \partial t} = 0, \quad \Delta \frac{\partial^2 \Psi}{\partial y \, \partial t} = 0, \quad \Delta \frac{\partial^2 \Psi}{\partial z \, \partial t} = 0.$$

[216] We can write the obtained results as

$$\Delta \frac{\partial}{\partial t} \left(\frac{\partial \Psi}{\partial x} + \frac{\partial F}{\partial t} \right) - K \mu \frac{\partial^3}{\partial t^3} \left(\frac{\partial \Psi}{\partial x} + \frac{\partial F}{\partial t} \right) = 0,$$

$$\Delta \frac{\partial}{\partial t} \left(\frac{\partial \Psi}{\partial y} + \frac{\partial G}{\partial t} \right) - K \mu \frac{\partial^3}{\partial t^3} \left(\frac{\partial \Psi}{\partial y} + \frac{\partial G}{\partial t} \right) = 0,$$

$$\Delta \frac{\partial}{\partial t} \left(\frac{\partial \Psi}{\partial z} + \frac{\partial H}{\partial t} \right) - K \mu \frac{\partial^3}{\partial t^3} \left(\frac{\partial \Psi}{\partial z} + \frac{\partial H}{\partial t} \right) = 0$$

or, in virtue of equalities (8.12),

$$\begin{cases} \Delta \overline{u} - K \mu \dfrac{\partial^2 \overline{u}}{\partial t^2} = 0, \\[2mm] \Delta \overline{v} - K \mu \dfrac{\partial^2 \overline{v}}{\partial t^2} = 0, \\[2mm] \Delta \overline{w} - K \mu \dfrac{\partial^2 \overline{w}}{\partial t^2} = 0. \end{cases} \tag{8.35}$$

Moreover, in virtue of equality (6.25), in a non-conductive medium where

$$u = 0, \quad v = 0, \quad w = 0,$$

the components $\overline{u}, \overline{v}, \overline{w}$ of the displacement current satisfy the equality

$$\frac{\partial \overline{u}}{\partial x} + \frac{\partial \overline{y}}{\partial v} + \frac{\partial \overline{w}}{\partial z} = 0. \tag{8.36}$$

Thus, in a non-conductive medium, the displacement currents are of transverse currents that propagate with the speed

$$V = \sqrt{\frac{1}{K\mu}}. \tag{8.29}$$

The preceding analysis is based on the use of equalities (6.19) and (6.25), natural consequences of Maxwell's *third electrostatics*; [217] one could therefore expect to encounter it in the memoir: *A Dynamical Theory of the Electromagnetic Field*; it is replaced there by another analysis that would be less easy to render accurate.

Maxwell ascertained a function χ, analogous to the function χ given by equality (7.82), which he later considered in his *Treatise*, such that

$$\Delta\chi = J. \tag{8.37}$$

He puts

$$\begin{cases} F = F' + \dfrac{\partial \chi}{\partial x}, \\[2mm] G = G' + \dfrac{\partial \chi}{\partial y}, \\[2mm] H = H' + \dfrac{\partial \chi}{\partial z}. \end{cases} \tag{7.83}$$

Equalities (7.55b), (8.37), and (7.83) obviously give

$$\frac{\partial F'}{\partial x} + \frac{\partial G'}{\partial y} + \frac{\partial H'}{\partial z} = 0, \tag{8.38}$$

such that F', G', H' can be regarded as the *transverse component of the electrotonic state* of which F, G, H are the components.

With equalities (7.83), equalities (8.26) become

$$\Delta F' - K\mu \frac{\partial^2 F'}{\partial t^2} = K\mu \frac{\partial}{\partial x}\left(\frac{\partial \Psi}{\partial t} + \frac{\partial^2 \chi}{\partial t^2}\right),$$

$$\Delta G' - K\mu \frac{\partial^2 G'}{\partial t^2} = K\mu \frac{\partial}{\partial y}\left(\frac{\partial \Psi}{\partial t} + \frac{\partial^2 \chi}{\partial t^2}\right), \tag{8.39}$$

$$\Delta H' - K\mu \frac{\partial^2 H'}{\partial t^2} = K\mu \frac{\partial}{\partial z}\left(\frac{\partial \Psi}{\partial t} + \frac{\partial^2 \chi}{\partial t^2}\right).$$

[218] Differentiating these equalities with respect to x, y and z, respectively, adding the obtained results member by member, and taking equality (8.38) into account; we find

$$\Delta\left(\frac{\partial\Psi}{\partial t} + \frac{\partial^2\chi}{\partial t^2}\right) = 0. \tag{8.40}$$

Maxwell does this calculation[53]; but instead of concluding with equality (8.40), he concludes, and this is not legitimate, with the equality

$$\frac{\partial\Psi}{\partial t} + \frac{\partial^2\chi}{\partial t^2} = 0. \tag{8.41}$$

Making use of this equality (8.41), he transforms equalities (8.39) into

$$\begin{cases} \Delta F' - K\mu\dfrac{\partial^2 F'}{\partial t^2} = 0, \\[2mm] \Delta G' - K\mu\dfrac{\partial^2 G'}{\partial t^2} = 0, \\[2mm] \Delta H' - K\mu\dfrac{\partial^2 H'}{\partial t^2} = 0. \end{cases} \tag{8.42}$$

The transverse part of the electrotonic state propagates with speed

$$V = \sqrt{\frac{1}{K\mu}}. \tag{8.29}$$

Moreover, equalities (8.37) and (8.41) give

$$\frac{\partial}{\partial t}\Delta\Psi + \frac{\partial^2 J}{\partial t^2} = 0,$$

and as $\Delta\Psi$ is, in a homogeneous medium, proportional [1st Part, equalities (4.16) and (4.16b)] to the density of the free electricity, [219] $\frac{\partial^2 J}{\partial t^2}$ is found to be proportional to $\frac{\partial e}{\partial t}$. Maxwell said[54]: "Since the medium is a perfect insulator, e, the free electricity, is immoveable;" this assertion does not logically follow from the electrostatics admitted in the memoir: *A Dynamical Theory of the Electromagnetic Field*. Maxwell has to admit, however, that $\frac{\partial^2 J}{\partial t^2}$ is necessarily null and conclude that a periodic electrical disturbance corresponds to a value in J different from 0.

The second electrostatics of Maxwell does not lend itself as well to the development of the electromagnetic theory of light as the third electrostatics of the same author.

---

[53] J. Clerk Maxwell, SCIENTIFIC PAPERS, vol. I, p. 581, equality (77).
[54] J. Clerk Maxwell, SCIENTIFIC PAPERS, vol. I, p. 582.

There are two points on which[55] all the electrostatics of Maxwell agree.

In the first place, two electric charges q_1, q_2, placed at a distance r from another inside a certain dielectric 1, repel with a force [1st Part, equalities (4.37) and (4.42)]

$$F = \frac{1}{K_1} \frac{q_1 q_2}{r^2}.$$

Secondly, a flat capacitor whose plates of area S are separated by a thickness θ of the same dielectric has a capacitance [1st Part, equality (4.46)]

$$C = \frac{K_1}{4\pi} \frac{S}{\theta}.$$

These two equalities, together with equality (8.29), immediately give these two laws, already obtained by Maxwell in his memoir: *On Physical Lines of Force* [220]:

1st LAW. *In a vacuum, the transverse displacement currents propagate with the same speed as light.*

2nd LAW. The specific inductive capacity compared to the vacuum is related to propagation velocities V_1 and V of the transverse displacement currents in the dielectric and in the vacuum, and to the magnetic permeability μ_1 of the dielectric by the relation

$$D_1 = \frac{V^2}{V_1^2} \frac{1}{\mu_1}. \tag{8.24}$$

These are the two essential laws of the electromagnetic theory of light. [221]

---

[55]To recognize this agreement, it must be remembered that the same quantity is named K in the *Treatise on Electricity and Magnetism* and here, $\frac{1}{E^2}$ in the memoir: *On Physical Lines of Force*, and $\frac{4\pi}{K}$ in the memoir: *A Dynamical Theory of the Electromagnetic Field*.

Chapter 9
Conclusion

The electromagnetic theory of light connects in such a fortunate way two disciplines hitherto distinct; it so fully meets the need, often manifested by physicists, to reconcile optical and electrical doctrines, that few people today hold it as invalid.

On the other hand, unless being blinded by a biased admiration, we cannot ignore the illogicalities and inconsistencies which render the reasonings of Maxwell unacceptable to an equitable spirit. These illogicalities, these inconsistencies, are also not, in the work of an English physicist, defects of minimal importance that are easy to correct; many illustrious geometers have sought to bring order into this work and had to give up.

Which side should we take, since we cannot resolve whether to accord a demonstrative value to the reasonings of Maxwell or to abandon the electromagnetic theory of light?

Many physicists today now lean toward the side that has been adopted by O. Heaviside,[1] by Hertz,[2] and by Cohn,[3] [222] of which Hertz[4] has clearly formulated the principle and claimed its legitimacy:

Since the reasonings and the calculations by which Maxwell developed his theory of electricity and magnetism are constantly undermined by non-accidental contradictions, not easy to correct, but essential and inseparable from the body of the work, let us leave aside these arguments and calculations. We simply take the equations to

---

[1]O. Heaviside. *On the Electromagnetic Wave-Surface* (PHILOSOPHICAL MAGAZINE, 5th series vol. XIX, p. 397; 1885.—HEAVISIDE's Electrical Papers, vol. II, p. 8).—*On Electromagnetic Waves, Especially in Relation to the Vorticity of the Impressed Forces and the Forced Vibrations of Electromagnetic Systems* (PHILOSOPHICAL MAGAZINE, 5th series, vol. XXV, p. 130; 1888.—ELECTRICAL PAPERS, vol. II, p. 375).

[2]H. Hertz. *Ueber die Grundgleichungen der Elektrodynamik für ruhende Körper* (WIEDEMANN'S ANNALEN. Bd. XL, p. 577; 1890.—UNTERSUCHUNGEN ÜBER DIE AUSBREITUNG DER ELEKTRISCHEN KRAFT, p. 208; 1894).

[3]Cohn. *Zur Systematik der Elektricitätslehre* (Wiedemann's Annalen, Bd. XL, p. 625; 1890).

[4]H. Hertz. UNTERSUCHUNGEN ÜBER DIE AUSBREITUNG DER ELEKTRISCHEN KRAFT: EINLEITENDE UEBERSICHT [English translation: Hertz (1893)], p. 21.

© Springer International Publishing Switzerland 2015   171
P.M.M. Duhem, *The Electric Theories of J. Clerk Maxwell*,
Boston Studies in the Philosophy and History of Science 314,
DOI 10.1007/978-3-319-18515-6_9

which they led Maxwell and, regardless of the processes by which these equations have been obtained, accept them as fundamental, as postulates on which we rest the whole edifice of the electrical theory. We will also keep, if not all the thoughts that have agitated the spirit of Maxwell, at least everything that is essential and indestructible in these thoughts, because "what is essential in the theories of Maxwell are his equations."[5]

Do we have right to set aside both the old electric theories and new theories by which Maxwell arrived at these equations and purely and simply to take these equations as the starting point for a new doctrine?

An algebraist always has the right to take a any group of equations and combine them according to the rules of calculation. The letters that certain relationships have linked will be involved in other relationships algebraically equivalent to the first.

But a physicist is not an algebraist. An equation is not simply, for him, the letters; these letters represent physical quantities which must be measurable experimentally or formed from other measurable quantities. Thus, if we are content to give a physicist an equation, it does not teach him anything; he must, to this equation, attach an indication of the rules by which he will make the letters of the equation correspond to the physical values they represent. Now, these rules, which cause him to know, [223] are all the hypotheses and reasonings by which he reached the equations in question; this is the *theory* that these equations summarize in symbolic form: *in physics, an equation, detached from the theory which led it there, makes no sense.*

According to H. Hertz, theories are identical when they lead to the same equations.

To the question[6]: "What is Maxwell's theory?" I do not know any shorter and more precise answer than this: "Maxwellz's theory is Maxwell's system of equations." Any theory which leads to the same equations, and, therefore, embraces the same set of possible phenomena, I will regard a form or particular case of Maxwell's theory; any theory which leads to other equations and, therefore, provides for the possibility of other phenomena, will be for me another theory.

This criterion is not sufficient to judge the equivalence of two theories. To be equivalent, it is not sufficient that the equations that they propose be literally identical; it is also necessary that the letters contained in these equations represent quantities related in the same way to measurable quantities, and to ensure this last characteristic, it does not suffice to compare the equations. We must compare the reasonings and hypotheses that constitute both theories.

So, one cannot adopt Maxwell's equations unless he arrives at them as a consequence of a theory of electric and magnetic phenomena; and since these equations are not consistent with the work of Poisson from classical theory, he will be forced to reject this classical theory, to break with the traditional doctrine, and to create with new concepts, on new hypotheses, a new theory of electricity and magnetism.

This is what Boltzmann did. [224]

---

[5][H. Hertz. *ibid*. p. 21].

[6]H. Hertz, *Abhandlungen über die Ausbreitung der elektrischen Kraft. Einleitende Uebersicht*, p. 23.

In a book published from 1891 to 1893,[7] he attempted in a prodigious effort to forget the doctrines that the tradition and usage teach us and to build, using entirely new concepts, a system where the equations of Maxwell are linked together logically.

There is no denying, in fact, that this book establishes a perfect link between the various equations written by Maxwell in his *Treatise on Electricity and Magnetism*. The contradictions and paralogisms of which Maxwell was so pleased, too often, to sow along the way that leads to these equations were carefully removed. Is this to say that the theory thus constructed is not open to criticism and satisfies all the desires of physicists? There will be many. Thus, the electrostatics of L. Boltzmann is only Maxwell's third electrostatics. Like the latter, it does not appear to agree with the actions that the charged conductors exert on dielectrics; the magnetism, copied from the memoirs of Maxwell, seems unlikely to be identified with the fertile doctrines of Poisson, F. Neumann, W. Thomson, and G. Kirchhoff—doctrines that Maxwell himself included in his *Treatise*.

So if, to arrive logically at Maxwell's equations, we follow the methods proposed by L. Boltzmann, we are forced to give up in part the work of Poisson and his successors on the distribution of electricity and magnetism, i.e. one of the the the most accurate and most useful parts of mathematical physics.

On the other hand, to save these theories, must we renounce all the consequences of the doctrine of Maxwell, and, in particular, the most appealing of these consequences, the electromagnetic theory of light? As Poincaré noted somewhere, it would be difficult to resolve.

Trapped in this dilemma: either to abandon the traditional theory of the electric and magnetic distribution, or to renounce the electromagnetic theory of light, can physicists adopt a third stance? Can they imagine a doctrine where the old electrostatics and magnetism and the new doctrine of the propagation of electric actions within dielectric media would be reconciled?[225]

This doctrine exists. It is one of the most beautiful works of Helmholtz[8]; a natural extension of the doctrines of Poisson, Ampère, Weber, and Neumann, it leads logically from the principles laid down at the beginning of the XIXth century to the most attractive consequences of Maxwell's theories, from Coulomb's laws to the electromagnetic theory of light; without losing any of the recent conquests of electrical science, it restores the continuity of the tradition.

---

[7]L. Boltzmann, *Vorlesungen über Maxwell's Theorie der Elektricität und des Lichtes. I^e Theil: Ableitung der Grundgleichungen für ruhende, homogeneous, isotropic Körper.—IIe Theil: Verhältniss zur Fernwirkungs-theory; specielle Fälle der Elektrostatik, stationaren Strömung und Induction.* Leipzig, 1891–1893.

[8]Helmholtz. *Ueber die Bewegungsgleichungen der Elektrodynamik für ruhende leitende Körper* (BORCHARDT'S JOURNAL FÜR REINE UND ANGEWANDTE MATHEMATIK, Bd. LXXII, p. 57, 1870.—WISSENSCHAFTLICHE ABHANDLUNGEN, Bd. I, p. 543).

References

Ampère AM (2015) Mathematical theory of electrodynamic phenomena, uniquely derived from experiments. Michael D. Godfrey. https://docs.google.com/file/d/0BzMHTgCmyrNZSVFCZkpXNTMwQU0/edit

Ariew R (1984) The Duhem thesis. Br J Philos Sci 35(4):313–325. http://www.jstor.org/stable/687336

Ariew R (2011) Pierre Duhem. In: Zalta EN (ed) The Stanford encyclopedia of philosophy, Spring 2011 edn. Stanford University, Stanford. http://plato.stanford.edu/entries/duhem/

Ariew R, Barker P (1986) Duhem on Maxwell: a case-study in the interrelations of history of science and philosophy of science. In: PSA: Proceedings of the biennial meeting of the philosophy of science association, pp 145–156. http://www.jstor.org/stable/193116

Bordoni S (2012a) Taming complexity: Duhem's third pathway to Thermodynamics. Editrice Montefeltro, Urbino, Italy. https://bit.ly/1c4smQm

Bordoni S (2012b) Unearthing a buried memory: Duhem's third way to Thermodynamics. Part 1. Centaurus 54(2):124–147. doi:10.1111/j.1600-0498.2012.00257.x, http://onlinelibrary.wiley.com/doi/10.1111/j.1600-0498.2012.00257.x/abstract

Bordoni S (2012c) Unearthing a buried memory: Duhem's third way to Thermodynamics. Part 2. Centaurus 54(3):232–249. doi:10.1111/j.1600-0498.2012.00259.x, http://onlinelibrary.wiley.com/doi/10.1111/j.1600-0498.2012.00259.x/abstract

Bordoni S (2013a) Routes towards an abstract thermodynamics in the late nineteenth century. Eur Phys J H 38(5):617–660. http://epjh.epj.org/articles/epjh/abs/2013/05/h130028/h130028.html

Bordoni S (2013b) When historiography met epistemology: Duhem's early philosophy of science in context. max-planck-institute for the history of science, Berlin

Brenner A, Needham P, Stump DJ, Deltete R (2011) New perspectives on Pierre Duhem's the aim and structure of physical theory. Metascience 20(1):1–25. doi:10.1007/s11016-010-9467-3, http://link.springer.com/article/10.1007/s11016-010-9467-3

Brush SG, Hall NS (2003) The kinetic theory of gases: an anthology of classic papers with historical commentary. Imperial College Press, London; Distributed by World Scientific Publishing, River Edge. http://site.ebrary.com/id/10255810

Buchwald JZ (1994) The creation of scientific effects: Heinrich Hertz and Electric Waves. University of Chicago Press, Chicago. http://site.ebrary.com/id/10448747

Cahan D (1993) Hermann von Helmholtz and the foundations of nineteenth-century science. University of California Press, Berkeley. http://search.ebscohost.com/login.aspx?direct=true&scope=site&db=nlebk&db=nlabk&AN=19172

© Springer International Publishing Switzerland 2015

P.M.M. Duhem, *The Electric Theories of J. Clerk Maxwell*,

Boston Studies in the Philosophy and History of Science 314,

DOI 10.1007/978-3-319-18515-6

Chervel A (2011) Lauréats des concours nationaux (1830–1950). http://www.inrp.fr/she/chervel_
    laureats1.htm#l1830
de Moura S, Sarmento IS (2013) Demechanization of electromagnetism and light pres-
    sure. Am J Electromagn Appl 1(3):53–61. doi:10.11648/j.ajea.20130103.12, http://article.
    sciencepublishinggroup.com/pdf/10.11648.j.ajea.20130103.12.pdf
Duhem H (1936) Un savant français. Pierre Duhem, Plon
Duhem PMM (1886) Le potentiel thermodynamique et ses applications à la mécanique chimique
    et à l'étude des phénomènes électriques. A. Hermann, Paris. http://gallica.bnf.fr/ark:/12148/
    bpt6k62445r
Duhem PMM (1888) De l'aimantation par influence. Annales de la faculté des sciences de
    Toulouse Mathématiques 2:1–138. doi:10.5802/afst.26, http://afst.cedram.org/item?id=AFST_
    1888_1_2__L1_0
Duhem PMM (1891–1892) Leçons sur l'électricité et le magnétisme. Gauthier-Villars, Paris. https://
    bit.ly/1hcRG5b
Duhem PMM (1893) Physique et métaphysique. Revue de questions scientifiques 34:55–83
Duhem PMM (1895a) Quelques remarques au sujet de l'électrodynamique des corps diélectriques
    proposée par J. Clerk Maxwell. In: Compte rendu du troisième Congrès scientifique international
    des catholiques tenu à Bruxelles du 3 au 8 septembre 1894, Congrès scientifique international
    des catholiques, pp 246–69, 337–8. https://books.google.com/books?id=1aOCAAAAIAAJ
Duhem PMM (1895b) Sur la Pression Dans les Milieux Diélectriques ou Magnétiques. Am J Math
    17(2):117–167. doi:10.2307/2369526, http://www.jstor.org/stable/2369526
Duhem PMM (1902) Les théories électriques de J. Clerk Maxwell: Étude historique et critique. A.
    Hermann, Paris. https://archive.org/details/lesthoriesle00duheuoft
Duhem PMM (1906) La théorie physique: son objet, et sa structure. Chevalier & Rivière, Paris.
    http://www.ac-nancy-metz.fr/enseign/philo/textesph/Duhem_theorie_physique.pdf
Duhem PMM (1908) ΣΩZEIN TA ΦAINOMENA: essai sur la notion de théorie physique de Platon
    à Galilée. A. Hermann, Paris. https://books.google.com/books?id=Wl43AQAAMAAJ
Duhem PMM (1911) Traité d'énergétique ou de thermodynamique générale. Gauthier-Villars, Paris.
    http://catalog.hathitrust.org/api/volumes/oclc/7963096.html
Duhem PMM (1913) Notice sur les Titres et Travaux scientifiques de Pierre Duhem rédigée par
    lui-même lors de sa candidature à l'Académie des sciences. No. 7 in Mémoires de la société
    des Sciences Physiques et Naturelles de Bordeaux, Gauthier-Villars, Paris. http://digital.ub.uni-
    duesseldorf.de/urn/urn:nbn:de:hbz:061:1-150836
Duhem PMM (1913–1959) Le système du monde: histoire des doctrines cosmologiques de Platon
    à Copernic. A. Hermann, Paris
Duhem PMM (1969) To Save the phenomena: an essay on the idea of physical theory from Plato
    to Galileo. University of Chicago Press, Chicago. http://catalog.hathitrust.org/api/volumes/oclc/
    45693.html
Duhem PMM (1985) Medieval cosmology: theories of infinity, place, time, void, and the plurality
    of worlds. University of Chicago Press, Chicago. http://site.ebrary.com/id/10462228
Duhem PMM (1990a) Logical examination of physical theory. Synthese 83(2):183–188. doi:10.
    1007/BF00413755, http://link.springer.com/article/10.1007/BF00413755
Duhem PMM (1990b) Research on the history of physical theories. Synthese 83(2):189–200. doi:10.
    1007/BF00413756, http://link.springer.com/article/10.1007/BF00413756
Duhem PMM (1991) The aim and structure of physical theory. Princeton University Press, Princeton.
    https://books.google.com/books?id=5mVPK7QBdTkC
Duhem PMM (1996) Essays in the history and philosophy of science. Hackett Publishing Company,
    Indianapolis. https://books.google.com/books?id=UofBybolmREC
Duhem PMM (2011) Commentary on the principles of Thermodynamics, Boston studies in the
    philosophy of science, vol 277. Springer, Netherlands. http://link.springer.com/book/10.1007/
    978-94-007-0311-7

Graneau P, Assis AKT (1994) Kirchhoff on the motion of electricity in conductors. Apeiron 19:19–25. http://freenrg.info/Scientific_Books/Kirchhoff_on_the_Motion_of_Electricity_in_Conductors.pdf

Hannam J (2009) God's philosophers: how the medieval world laid the foundations of modern science. Icon Books, London

Hertz H (1893) Electric waves being researches on the propagation of electric action with finite velocity through space. Macmillan, London

Howard D (1990) Einstein and Duhem. Synthese 83(3):363–384. doi:10.1007/BF00413422, http://link.springer.com/article/10.1007/BF00413422

Jaki SL (1984) Uneasy genius: the life and work of Pierre Duhem. The Hague, Boston. http://link.springer.com/book/10.1007/978-94-009-3623-2/

Kirchhoff G (1857a) LIV. On the motion of electricity in wires. Philos Mag Ser 4 13(88):393–412. doi:10.1080/14786445708642318, http://www.tandfonline.com/doi/abs/10.1080/14786445708642318

Kirchhoff G (1857b) Ueber die Bewegung der Elektricität in Drähten. Annalen der Physik 176(2):193–217. doi:10.1002/andp.18571760203, http://onlinelibrary.wiley.com/doi/10.1002/andp.18571760203/abstract

Kragh H (2008) Pierre Duhem, entropy, and Christian faith. Phys Perspect 10(4):379–395. doi:10.1007/s00016-007-0365-z, http://link.springer.com/article/10.1007/s00016-007-0365-z

Lorentz HA (1926) Maxwells elektromagnetische theorie. In: Sommerfeld A (ed) Encyklopädie der mathematischen Wissenschaften mit Einschluss ihrer Anwendungen, Physik, vol 5. Verlag und Druck von B. G. Teubner, Leipzig, pp 140–141. http://gdz.sub.uni-goettingen.de/dms/load/toc/?PPN=PPN360709672

Lorenz L (1867) XXXVIII. On the identity of the vibrations of light with electrical currents. Philos Mag Ser 4 34(230):287–301. doi:10.1080/14786446708639882, http://www.tandfonline.com/doi/abs/10.1080/14786446708639882

Maiocchi R (1985) Chimica e filosofia: scienza, epistemologia, storia e religione nell'opera di Pierre Duhem. La Nuova Italia, Firenze

Maugin GA (2014) Helmholtz interpreted and applied by Duhem. In: Continuum mechanics through the eighteenth and nineteenth centuries, no. 214 in solid mechanics and its applications, Springer International Publishing, pp 99–112. http://link.springer.com/chapter/10.1007/978-3-319-05374-5_7

Miller DG (1970) Duhem, Pierre Maurice Marie. http://www.encyclopedia.com/topic/Pierre_Maurice_Marie_Duhem.aspx#1

Needham P (2013) Unearthing a buried memory. Metascience 1–5. doi:10.1007/s11016-013-9832-0, http://link.springer.com/article/10.1007/s11016-013-9832-0

Neumann C (1868) Theoria nova phaenomenis electricis applicanda. Annali di Matematica Pura ed Applicata 2(1):120–128. doi:10.1007/BF02419606, http://link.springer.com/article/10.1007/BF02419606

O'Rahilly A (1938) Electromagnetic theory: a critical examination of fundamentals. Longman's, Green and Co., Cork. https://archive.org/details/ElectrodynamicsORahilly

Oxford English Dictionary (2014) mile, n.1. http://www.oed.com/view/Entry/118382

Parkinson EM (2008) Rankine, William John Macquorn. http://www.encyclopedia.com/topic/William_John_Macquorn_Rankine.aspx#1

Pascal B (2004) Pensées. Hackett Publishing Company Inc., Indianapolis

Rankine WJM (1855) Outlines of the science of energetics. Proc Roy Philos Soc Glasg 3:121–141. https://en.wikisource.org/wiki/Outlines_of_the_Science_of_Energetics

Riemann B (1867) Ein Beitrag zur Elektrodynamik. Ann Phys 207(6):237–243. doi:10.1002/andp.18672070605, http://onlinelibrary.wiley.com/doi/10.1002/andp.18672070605/abstract

Roy L (1915) L'électrodynamique de Helmholtz-Duhem et son application au problème du mur et à la décharge d'un condensateur sur son propre diélectrique. Annales de la faculté des sciences de Toulouse Mathématiques 7:221–245. doi:10.5802/afst.302, http://afst.cedram.org/item?id=AFST_1915_3_7_221_0

Roy L (1918) L'électrodynamique de Helmholtz-Duhem et son application au problème du mur et à la décharge d'un condensateur sur son propre diélectrique (Suite). Annales de la faculté des sciences de Toulouse Mathématiques 10:1–63. doi:10.5802/afst.306, http://afst.cedram.org/item?id=AFST_1918_3_10_1_0

Roy L (1923a) L'électrodynamique des milieux isotropes en repos d'après Helmholtz et Duhem. Gauthier-Villars, Paris. https://archive.org/details/LelectrodynamiqueDesMilieuxIsotropesEnReposDapresHelmholtzEtDuhem

Roy L (1923b) Sur l'électrodynamique des milieux en mouvement. Annales de la faculté des sciences de Toulouse Mathématiques 15:199–241. doi:10.5802/afst.329, http://afst.cedram.org/item?id=AFST_1923_3_15_199_0

Weber W (1893) Wilhelm Weber's Werke, vol 3. Springer, Berlin. http://link.springer.com/book/10.1007/978-3-662-24693-1

Weber W (1894) Elektrodynamische Maassbestimmungen insbesondere über elektrische Schwingungen. Wilhelm Weber's Werke. Springer, Berlin, pp 105–241. http://link.springer.com/chapter/10.1007/978-3-662-24694-8_5

Weber W (2007) Determinations of electrodynamic measure: concerning a universal law of electrical action. 21st century science & technology, Leesburg. http://www.21stcenturysciencetech.com/Articles2007/Weber_1846.pdf

Weber W, Kohlrauch R (2003) On the amount of electricity which flows through the cross-section of the circuit in galvanic currents. In: Bevilacqua F, Giannetto EA, Hecht L (eds) Volta and the history of electricity, Università degli studi di Pavia and Editore Ulrico Hoepli, Milano, pp 267–286. http://www.ifi.unicamp.br/assis/Weber-Kohlrausch(2003).pdf

Wipf N (2011) Pierre Duhem (1861–1916) et la théorie du magnétisme fondée sur la thermodynamique. Ph.D. dissertation, Université Lille 1, http://ori.univ-lille1.fr/notice/view/univ-lille1-ori-21798

Index

© Springer International Publishing Switzerland 2015
P.M.M. Duhem, *The Electric Theories of J. Clerk Maxwell*,
Boston Studies in the Philosophy and History of Science 314,
DOI 10.1007/978-3-319-18515-6

Printed in the United States
By Bookmasters